Reasons Why

Reasons Why

Bradford Skow

OXFORD
UNIVERSITY PRESS

Great Clarendon Street, Oxford, OX2 6DP,
United Kingdom

Oxford University Press is a department of the University of Oxford.
It furthers the University's objective of excellence in research, scholarship,
and education by publishing worldwide. Oxford is a registered trade mark of
Oxford University Press in the UK and in certain other countries

First Edition published in 2016

Impression: 1

Published in the United States of America by Oxford University Press
198 Madison Avenue, New York, NY 10016, United States of America

British Library Cataloguing in Publication Data

Data available

Library of Congress Control Number: 2016943966

ISBN 978-0-19-878584-2

Printed in Great Britain by
Clays Ltd, St Ives plc

For Elliot and Nathaniel
Don't stop asking

Contents

Acknowledgments

For reading parts of the manuscript, and/or providing valuable feedback and advice, thanks to Josh Schechter, the students in his spring 2015 seminar, Quinn White, Nathaniel Baron-Schmitt, Daniel Muñoz, Nina Emery, Steve Yablo, Philip Kitcher, Kieran Setiya, Jack Spencer, Justin Khoo, Greg Frost-Arnold, Zoltán Szabó, Jason Stanley, Shelly Kagan, and Shamik Dasgupta. An extra helping of gratitude to Jonathan Shaheen, who read the entire manuscript and wrote over 8000 words in response. Thanks also to audiences at: MIT, Yale, Columbia, Brown, the 2013 "Explaining without Causes" conference in Cologne, the 2014 "Explanation Beyond Causation" conference at LMU-Munich, the 2014 New York Philosophical Association meeting in Syracuse, and the 2015 Pacific APA meeting in Vancouver. The writing of this book was made possible by Taza® Chocolate.

Lyrics from *Rattling Locks*, composed and written by Josh Ritter, reprinted by permission of Rural Songs.

Frostbite and heartsickness
ain't neither one of them so bad if you can understand
the reason why

Josh Ritter, "Rattling Locks"

1

A Few Opening Remarks

I was tempted to subtitle this book "a theory of explanation," so casual browsers running across it in their local bookstore would have a better idea what it was about. But that would betray everything the book stands for. One of my main claims is that so-called "theories of explanation" are not really theories of explanation at all—or, at least, they shouldn't be theories of explanation. What they are, or should be, are theories of answers to why-questions.

The subject matter of such theories is interesting for its own sake—we are all born wanting to know why, and, if things go right, we never stop. The things we ask about range from the mundane to the esoteric: why does my brother keep hitting me?; why are the stars visible only at night?; why are we here at all? Given the importance of why-questions to our cognitive lives, a philosopher naturally wants to know what it takes to be an answer to a why-question.

Besides being themselves a topic for philosophical investigation, why-questions and their answers come up in philosophical investigations of other topics. Why-questions are important, for example, in the philosophy of science. "Scientific realism" means different things to different people, but one idea commonly associated with this doctrine is that it is one of the aims of science to answer why-questions. Science aims to figure out why things happen, not just describe, even very systematically, what happens.

Why-questions are also important in the philosophy of action. They help identify one central topic in the philosophy of action, namely intentional action: someone acts intentionally if, and only if, one may ask "Why did he do it?"—where "why" is used in a special way, different from the way it is used when we ask why the moon is waning.[1]

[1] This is Anscombe's proposal in *Intention* (§5).

Why-questions are important in metaphysics. The obvious example is their importance to the theory of grounding. Many metaphysicians are very busy producing theories of grounding—but what *is* grounding? A common strategy for helping initiates get a handle on the subject matter of these theories is to say that when one fact grounds another, the first may be used to answer the question why the second obtains. Another example of a part of metaphysics where why-questions are important is the theory of modality. If some fact F obtains in two possible worlds W and V, then those worlds are in one respect similar. Boris Kment argues, in "Counterfactuals and Explanation," that this respect of similarity matters for how close V and W are, in the sense of closeness relevant to evaluating counterfactuals, if and only if the question why F obtains has the same answer in both worlds.

Why-questions are important in normative ethics. When evaluating a theory of right action, utilitarianism for example, it is not enough to check whether it correctly sorts acts into right and wrong—whether every (possible) right act is classified as right by the theory, and every (possible) wrong act classified as wrong. A theory that correctly sorts acts into right and wrong is still false if it gives the wrong answer to the question of why right acts are right.

Why-questions are important in the philosophy of mind. The most appealing—to me—view about the status of mental facts is physicalism, the view that all mental facts are determined by physical facts. But there is a well-known problem for physicalism, a problem known as the "explanatory gap." What is this problem? What is the gap? The explanatory gap is just this: physicalists have no answer to the question why the physical facts determine the mental ones in the way that they do.[2]

Why-questions of course are important in epistemology. "Inference to the best explanation" is the pattern of belief formation whereby, of the possible answers to the question why one's evidence obtains (or, perhaps, why one has that evidence), one comes to believe the "best" answer. Why-questions also appear in some theories of knowledge. Kieran Setiya defends the thesis that X knows p only if X formed the belief in p by a reliable method, and X used that method *because* it is reliable.[3] (That

---

[2] The term "explanatory gap" was coined by Joseph Levine, in "Materialism and Qualia."

[3] See *Knowing Right from Wrong*, pp. 96–9 for citations of other philosophers with similar views. Setiya's view is more subtle than is apparent from this one sentence sketch.

is, "because it is reliable" is the answer to the question why X used that method.)

When one surveys all the answers science has given to all the why-questions it has addressed, it may be as if one finds oneself in an island rainforest, marveling at the magnificent diversity of what there is to see. To deny that diversity, one might think, is to give in to a reductive impulse that should be resisted. The right response is to celebrate and catalog that diversity, as a biologist would, were she to run across a jungle teeming with unknown species of parrot.

This is not my view. I hold that, however different answers to why-questions look on the surface, deeper down there is a great deal of unity to them. Or, at least, that is my view about answers to why-questions about events. In this book I will (almost) defend a version of the idea that an answer to the question why some given event E happened must cite causes of E.[4] This idea has been around for a while, but (you will not be surprised to learn) I don't think others' defenses of it have been good enough. Witness the continuing production and publication of examples of "non-causal explanations."

My theory is organized around the "reason why." One may answer the question why Q by saying something of the form "one reason why Q is that R," and I think that a theory of answers to why-questions should take the reason why as its central object of analysis. It is the central object of analysis of my theory.

This is philosophy, so I will admit up front that my theory of reasons why may be false. While it would, to be honest, take a lot to get me to accept that it is false, it would take a whole lot more to get me to accept that it was a mistake to focus on reasons why in the first place. Put another way, philosophers rarely persuade each other of anything, and so I know that those who disagree with me are unlikely to be persuaded that my theory of reasons why is the right one. I do hope, however, to persuade them that a theory of answers to why-questions should be a theory of reasons why, even if it should not be the one I defend.

I've been using "reason why" as if it were a noun phrase, when it is not. "Reason why" is not a semantically complete unit at all; in this respect, and in many others, it is analogous to "person who," or "place where." It

---

[4] "Almost" because I also hold that an answer can cite a ground of E, in the metaphysicians' sense of "ground."

would make little sense to propose a theory of persons who, or of places where. People who *what*? Places where *what*? So to prevent readers from thinking that here, just a few pages in, my theory is already incoherent, let me remark that when I say that I will propose a theory of reasons why I mean that I will propose a theory of what it takes for one fact to be a reason why some other fact obtains. Having said this, I will often in this book use "reason why" as a noun phrase, for stylistic convenience.

Both to formulate my theory, and to defend it, I make use of a distinction, a distinction that is much easier to see when we think in terms of reasons why than if we don't. It is a distinction between different levels of reasons why. There are, on the one hand, the reasons why it is the case that F, and on the other, the reasons why those reasons are reasons. This distinction is an essential component of my strategy for explaining away many apparent counterexamples to the idea that answers to why-questions about events must cite causes.

But before I say anything about reasons why, much less different levels of reasons why, I need to say something about explanation: about why my theory of explanation isn't a theory of explanation, and about why no one else's should be either.

2

From Explanations to Why-Questions

2.1 Against Explanation

Why do philosophers of science use "theory of explanation" as a general term for theories of explanation? That sounds like a dumb question, but bear with me.

"The answer is obvious," you might say; "just look at some examples of the kind of thing philosophers of science have wanted a theory of explanation to account for." Okay, let's do that. Here are some examples. Scientists wanted, and then found, an explanation of the extinction of the dinosaurs: a comet or asteroid hit the Earth.[1] They wanted to explain the tides, and Newton finally did, when he developed his theory of universal gravitation. Scientists also wanted to explain the photoelectric effect. The photoelectric effect is when electrons fly off a metal when light is shone on it; the number of electrons that fly off, and their speed, depends on the character of the incident light. If you make the light brighter, more electrons fly off, but they do not fly off any faster; but if you increase the light's frequency—make it bluer—*then* the electrons fly off faster. Classical electromagnetism couldn't explain this; it predicted, falsely, that brighter light *will* make the electrons fly off faster. (Classically, a brighter light means a greater force on each electron, and a greater force makes for a greater final speed.) It was quantum mechanics that provided the correct explanation. Quantum mechanics says that light is made up of individual particles, or quanta, namely photons. Simplifying a lot, electrons fly off a metal when photons hit them. To make an electron fly off faster it has to be hit by a higher-energy photon. And the energy of a photon corresponds to the frequency of the light, not its intensity.

---

[1] At least, this a widely-accepted explanation.

The words "explain" and "explanation" are all over the last paragraph; together they occur five times. So what is the problem? Surely a theory that accounts for what is going on in these examples, a theory of the kind of scientific achievement they are instances of, deserves to be called a "theory of explanation."

I disagree. As a first step toward understanding why I disagree, ask yourself what in each case is "being explained." To make it concrete imagine to yourself a physicist, Sue, and a student Marcel; Marcel knows that electrons fly off metals, but doesn't know the correct theoretical account of this effect. Sue imparts to Marcel the knowledge I sketched above, in a way that makes it true that

(1)   Sue explained the photoelectric effect to Marcel.

In (1) the object of the verb "explained" is "the photoelectric effect"—so isn't it the photoelectric effect that is being explained?

Not really. The grammar here is misleading. Think for a minute about epistemology. Epistemologists aim to understand knowledge. But that is not quite right. What epistemologists aim to understand is the *propositional attitude* knowledge—what we ascribe when we say things like "Smith knows that Barack Obama was born in Hawaii," or "Bloggs knows where the President lives."[2] But not all uses of "know" attribute a propositional attitude. One might say, for example, that Fred knows Ed, meaning not that ⌜Fred knows that P⌝ is true for some sentence P that is about Ed, but that Fred has met Ed, that they are friends, or something like that.[3] Now suppose I present a certified epistemologist with the sentence

(2)   John knows Bill's telephone number.

and ask her for the conditions under which (2) is true. On the surface (2) looks to have the same form as "Fred knows Ed," so the epistemologist might say that answering my question falls outside her domain of expertise. But this would be a mistake. Despite their surface similarity, "knows"

---

[2] A standard view is that the first sentence attributes a relation between Smith and the proposition that Barack Obama was born in Hawaii, and the second attributes a relation between Bloggs and the proposition that is the answer to the question of where the President lives. (Which proposition this is will depend on context; in one context it may be the proposition that the President lives in Washington, D.C.)

[3] This is not meant to be or imply any particular analysis of "X knows Y" where "Y" holds the place for a singular term or description, rather than a that-clause.

in (2) is not being used in the same way as in "Fred knows Ed." For it is evident that (2) is equivalent to

(3)   John knows *what Bill's telephone number is*,

and in (3) "knows" is being used in its propositional-attitude sense.

Linguists say that when a noun phrase like "Bill's telephone number" is used so that it has the same meaning as an indirect question (here, "what Bill's telephone number is"), it is being used as a "concealed question."[4] Just as it is evident that "Bill's telephone number" is a concealed question in (2), it is evident that "the photoelectric effect" is a concealed question in (1). Make the question explicit and we get

(4)   Sue explained to Marcel why the photoelectric effect happens (why electrons fly off metals when light is shone on them).[5]

Insofar as philosophers of science are after what is going on when sentences like (1) are true, they are really after what is going on when sentences like (4) are true. And (4) contains "why" as well as "explains."

This fact raises the following question: while philosophers of science call the kind of theory they are after a theory of explanation, is there any reason to think that the phenomena of interest involve explanation in any essential way? Maybe what they are after is a theory of why-questions, not a theory of explanation.

Enough coyness. These hypotheses are true. The phenomena of interest to philosophers of science do *not* involve explanation in any essential way. A theory of why-questions *is* what they are after.

One fact that favors my thesis is that explanation and why-questions come apart, and the interests of philosophers of science seem to go with why-questions, not with explanation. One way in which explanation and why-questions come apart is that not all cases of explaining are cases of explaining why. The verb "explain" can take any indirect wh-question as

---

[4] The earliest reference to concealed questions in the linguistics literature that I have found is to Baker's *Indirect Questions in English*. See also Heim, "Concealed Questions"— the (2)–(3) example is hers. Sylvain Bromberger was aware of concealed questions when he was working on theories of explanation in 1962, though they had not yet been given that name (see "An Approach to Explanation" p. 45).

[5] A noun phrase may conceal different questions in different contexts. There are contexts in which (1) is equivalent to "Sue explained what the photoelectric effect is." I am assuming that we are not in a context like that.

a complement, not just a why-question.[6] Someone can explain *where* the bread is, *who* is coming to the party, or *how* to ride a bike. But figuring out what was going on when John explained who was coming to the party does not seem like a job for the philosophy of science. (The same goes for explaining where, or explaining what; the case of explaining how is more controversial.)

This observation is consistent with a weaker thesis than mine. Maybe philosophers of science are not interested in explanation-in-general; it could still be that they are interested in explaining why.

To explain is to explain the answer. To explain who is coming to the party is to explain the answer to the question who is coming to the party; to explain why the car skidded off the road is to explain the answer to the question why the car skidded off the road.[7] And there are two separable parts to explaining the answer to a wh-question. The first part is to—somehow—convey to one's audience what that answer is. One can do this first part in many ways; and it can be done without explaining anything. One can, for example, merely *tell* one's audience what the answer is.[8] The second part involves doing the first part in a way that counts as *explaining*. Maybe to explain an answer one must go through it slowly and clearly.[9]

In any regard, a theory of explanation, properly so-called, will say only what it takes for something to be an act of explaining. Saying what it

---

[6] Bromberger made this observation in the 1960s ("An Approach to Explanation," p. 20; see also "Why-Questions," p. 80), and G. A. Cohen made it in the 1970s (*Karl Marx's Theory of History*, pp. 251–2), but their observations did not succeed in turning philosophers' attention away from explanation. (Bromberger's papers did get Kitcher to acknowledge that his unificationist theory of explanation was a theory of explaining why, not a general theory of explanation ("Explanatory Unification," p. 510, note 3).)

A wh-question (in English) is a question asked using one of the interrogative words "who," "which," "what," "when," "where," "how," or "why." Bromberger claimed that there are some indirect wh-questions that sound odd as complements to "explained": he held that it sounds odd to say that John explained what time it is. If this is odd, I do not think it is because the sentence is ungrammatical.

"Explain" can also take a that-clause as a complement: "He said he couldn't be my penpal, because he already had one, so I explained to him that you can have more than one penpal."

[7] Probably "the answer" in "explain the answer" is another concealed question, taking the place of "what the answer is."

[8] This is another way in which explanation and why-questions come apart: one can *answer* the question why (say) the moon is rising without *explaining* why the moon is rising.

[9] Cohen suggests this (*Karl Marx's Theory of History*, p. 152, note 1). Bromberger held a different view, which I do not find plausible ("An Approach to Explanation").

takes to be an answer to a wh-question is the job of a different theory, or collection of theories. So if you want to know what it takes for someone to have explained the answer to a why-question, you will have to do more than consult your best theory of explanation. You will have to combine what that theory tells you about what it takes to *explain* the answer to a question with what your best theory of answers to why-questions says it takes to *be* the answer to a why-question.

It is obvious that philosophers of science should want a theory of answers to why-questions. The notion of an answer to a why-question plays an important role in science, and in our thinking about science. Scientific realism, on one interpretation of that slippery doctrine, includes the view that science aims to figure out why the phenomena are as they are, not just to describe those phenomena in some systematic way. A scientific theory is supposed to be better—more worthy of our belief—than its rivals to the extent that it, if true, answers more why-questions, or gives better answers to why-questions, than its rivals.

I do not see that philosophers of science should also want a theory of explanation properly so-called. Explaining is, primarily, a speech act, something people do with words.[10] We do talk of theories explaining things, but this is a derivative use. A theory explains when someone can use that theory to perform the speech act of explaining. A theory of explanation properly so-called will say what distinguishes this speech act from others. I don't deny that it is an interesting question what distinguishes this speech act, but if you had to assign the task of answering it to a branch of philosophy, you would assign it to the philosophy of language, not to the philosophy of science.

Imagine that scientists stopped explaining the answers to why-questions. They continued to seek and discover answers to why-questions, they continued to come to know the answers to why-questions, and pass that knowledge on to the general public (maybe they stopped doing this clearly and patiently). They wouldn't stop being scientists if they changed in this way, nor would they be worse scientists (they might of course be worse *public intellectuals* or *members of the political community*). I think that the questions raised by science for the philosophy of science in this

---

[10] J. L. Austin divided speech acts into five categories, and placed explaining in group 7b of the "expositives" category, a group that also includes illustrating and formulating (*How To Do Things With Words*, p. 163).

scenario are just the same as those raised by science as it is actually practiced. So a theory of the speech act of explaining is not part of the philosophy of science; a complete philosophy of science only needs a theory of why-questions.

This is not meant to be a silly exercise in policing the boundaries between the parts of philosophy. My point is that the impulse that has led philosophers of science to pursue a "theory of explanation" should have led them to pursue a theory of answers to why-questions. My point is that the cluster of questions philosophers of science want to answer, that they think a theory of explanation will answer, will be answered instead by a theory of why-questions.

Failure to appreciate these points has fostered confusion. Think about Carl Hempel's DN model of explanation.[11] The DN model says, roughly, that X explains Y iff X is a sound argument with Y as its conclusion that essentially contains a law-stating premise (a premise that expresses a law of nature). This model fits many examples in which someone explains *why* something is the case, but isn't even remotely plausible as a theory of what is happening when a policeman explains where the train station is, or when a biologist explains what the theory of evolution says. (Hempel emphasized that his was a theory of *scientific* explanation, so maybe the policeman example isn't fair, but in the second example we have a scientist explaining a scientific theory.) Michael Scriven cited an example like these as an objection to Hempel's theory, maintaining that the theory is hopeless because "it does not fit the case of explaining . . . the rules of Hanoverian succession."[12] Here "the rules of Hanoverian succession" is another concealed question; Scriven's objection is better put by saying that one need not produce any valid arguments or cite any laws of nature to explain what the rules of Hanoverian succession are.[13] If I were defending the DN model I would reply: *so what?* Hempel may have called the DN model a

---

[11] The DN model (sometimes known as the D-N model) is developed most fully in Hempel's famous essay "Aspects of Scientific Explanation." It will appear now and then in this book, playing a supporting role.

[12] Scriven's objection is in "Truisms as the Grounds for Historical Explanation," p. 452; the quote is from Hempel's statement of the objection ("Aspects," p. 413). See also Scriven's "Explanations, Predictions, and Laws," p. 177.

[13] Among other things, the rules, as set out in the Act of Settlement of 1701, made Sophia of Hanover and her Protestant descendants heirs to the throne of England.

model of "explanation," but he shouldn't have.[14] It is best evaluated as a theory of answers to why-questions. Understood as a theory of that kind, it says nothing about answers to what-questions, or about what it takes to explain an answer to a what-question.

But this is not Hempel's reply to Scriven. Hempel seems to have been committed to labeling his theory a theory of explanation, so he could not dismiss Scriven's objection on the grounds I cited. Instead, sounding a little irritated, Hempel complained that Scriven's objection "is like objecting to a metamathematical definition of proof on the ground that it does not fit the use of the word 'proof' in 'the proof of the pudding is in the eating', nor in '86 proof Scotch'" (p. 413). So Hempel's reply is that "explain" is ambiguous, that he means to focus on just one of the word's meanings, and that in Scriven's example "explain" is not being used with its target meaning. But there is no more reason to think that "explained" means different things in these two sentences:

(5) John explained what the rules of Hanoverian succession are.

(6) John explained why these are the rules of Hanoverian succession.[15]

than to think that "knows" means different things in these two:

(7) John knows what the rules of Hanoverian succession are.

(8) John knows why these are the rules of Hanoverian succession.

In the second pair of sentences "knows" means the same thing, and the difference in the sentence's meaning comes from the difference in meaning of the embedded question. The same goes for the first pair.

I think this is completely obvious, but I can also give an argument. If "knows" means different things in (7) and (8) we shouldn't be able to make sense of their "reduced conjunction," since the reduced conjunction has only one occurrence of "knows," and it can have only one meaning:

John knows what the rules are and why they are the rules.

---

[14] Bromberger in 1966: "As a general characterization of the notion of explanation, that is, as a description of the truth-conditions of statements of the form 'A explains B' . . . the Hempelian doctrine obviously will not do" ("Why-Questions," p. 82).

[15] Because as of 1700 Princess Anne was the only remaining legal heir to the throne.

Similarly, if "explained" means different things in (5) and (6) we shouldn't be able to make sense of

John explained what the rules are and why they are the rules.[16]

But the sentences are perfectly in order.

Hempel was wrong to say that "explains" is ambiguous. But he did not need to say this in order to respond to Scriven's objection.[17]

So: it is a theory of why-questions that we should be after. This is a claim about what philosophers of science should be after, not what they take themselves to be after, or what they have in fact pursued.[18] There are philosophers who have tried to produce theories of explanation that cover explaining how in addition to explaining why, and others who have spent a long time thinking about what distinguishes explaining from other speech acts.[19] Their theories are not relevant to the philosophy of science.

I will say a bit more in defense of this claim in the next section. But first: am I over-reacting? Sylvain Bromberger noted long ago that the noun "explanation" is ambiguous; it can name either an act of explaining, or the product of, the information conveyed by, an act of explaining

---

[16] Thanks here to Jonathan Shaheen.

[17] It is a sign of Hempel's long shadow that his false claim continues to be accepted. Salmon endorsed Hempel's reply to Scriven forty years after Hempel first gave it (Salmon, *Four Decades*, p. 6). And almost fifteen years after that James Woodward, in *Making Things Happen*, seems to accept the thesis that "explain" is ambiguous, or at least that it is used differently in "explained why" and "explained where" (p. 4). G. A. Cohen claimed that Hempel was wrong to say that "explains" is ambiguous in *Karl Marx's Theory of History* (p. 252, note 1), but his remark had no influence.

[18] Though I agree with Bromberger: "Many writings ostensibly about explanation are best understood as not about explanation at all but as about what is called for by the questions belonging to one or the other of these groups" of questions, why-questions included ("An Approach to Explanation," p. 19). He took this to be a mistake, writing that "Philosophers of science should be able to state the truth-conditions that govern . . . 'to explain'" (p. 18). Needless to say, I disagree.

Van Fraassen is probably the most notable philosopher of science of the last forty years who was explicitly after what I have said he should have been, namely a theory of why-questions (see chapter 5 of *The Scientific Image*; I will say more about his views later). But he also *identified* theories of explanation with theories of why-questions (see p. 134 of his book), which I think was a mistake (though see the paragraph following the one to which this note is attached). I will say something about Hempel's views on the relationship between explanations and why-questions presently.

[19] For example, Cross, "Explanation and the Theory of Questions"; Achinstein, *The Nature of Explanation*.

("An Approach to Explanation," p. 50). I've complained that the act of explaining is not something philosophers of science should be interested in. But what about the second thing? Well, I've also complained that not all explanations, in the sense of pieces of information conveyed by an act of explaining, are things that philosophers of science should be interested in, since they include pieces of information that constitute answers to how-questions and to where-questions in addition to answers to why-questions. Nevertheless, to be fair, it is true that sometimes, in some contexts, we reserve "explanation," as a general term for the product of acts of explaining, for the product of acts of explaining answers to why-questions. For an example of such a context, here is the journalist Andrew Gumbel writing in *The Guardian*:

> Perhaps the most striking thing about the Oklahoma City bombing—by far the most destructive act perpetrated by a home-grown assailant against fellow Americans—is not how much we've learned over the past 20 years but rather how much we still do not know.
>
> Despite the government's insistence that the case has been solved, we don't know the exact origin of the plot or how many people carried it out.... We don't know how McVeigh and Nichols learned to build a fertiliser bomb of such size and power.... We don't know the identities of the other people seen with McVeigh on the morning of the bombing.... There is no ready explanation for a different Ryder truck seen by witnesses at McVeigh's motel in Kansas and at the state park where the bomb was assembled in the week leading up to the bombing; no explanation for the other people seen inside McVeigh's motel room during the same period; no satisfactory explanation of the fact that two people were seen renting the bomb truck on 17 April, neither of them entirely fitting McVeigh's description.
>
> ("Oklahoma City Bombing: 20 years later, key questions remain unanswered")

Gumbel lists many questions that remain unanswered. Some are how-questions (how McVeigh and Nichols learned to build the bomb), some are who-questions (who were the people seen with McVeigh on the morning of the bombing), and some are why-questions. He applies "explanation" only to the (unknown) answers to the why-questions. Not only does he do this, he finds it more natural to talk of explanations than to talk of why-questions and their (unknown) answers. He could have gone from "We don't know how McVeigh and Nichols . . . " and "We don't know the identities . . . " to "We don't know why there was a different Ryder truck at McVeigh's motel"—certainly that would have preserved a parallelism of structure. But instead he went with "There is no ready explanation for a different Ryder truck."

Of course, "explanation" is not in every context used only for answers to why-questions. But the article from *The Guardian* shows that in some contexts it clearly is, and maybe contexts like that are more common than not. Why not assume that philosophers debating the nature of explanation have been in contexts like that?[20] Isn't it harmless to put yourself in one of these contexts, and while you're there call your theory a theory of explanation?

No. Calling your theory a theory of explanation leads to thinking that your theory should be a completion of

(9)   Fact F is an explanation of fact G iff . . .

But choosing to put your theory in the form (9) is far from a harmless choice. For one thing, as I have already said—but this is a minor point—focus on the word "explanation" has led some philosophers of science to investigate topics that have nothing to do with what it takes to be an answer to a why-question. More importantly, even if we agree right now to use "explanation" for answers to why-questions and nothing more in our philosophical writing, we make sure everyone knows we've made this agreement, and we charitably read this use back into the writings of others, we will still be left blind to important distinctions. There are distinctions to be made—and that I will make—between partial answers and complete answers to why-questions, and between "merely" partial answers and partial answers that are also *part* of the complete answer to a why-question. How are we to interpret a theory that takes the form (9)? Is it a theory of partial answers, of complete answers, of parts of complete

[20] Charity certainly suggests that we make this assumption in many cases. For example, charity suggests that we make this assumption about the context Larry Wright was in when he wrote his book *Teleological Explanations*; otherwise we would have to fault him for falsely equating explanations with answers to why-questions, as he does in this passage: "he is driven to this formulation by a curious and unfortunately widespread prejudice about the nature of explanation, namely, *that the explanation of P, the answer to the question 'Why did P occur?', must itself rule out all alternatives to P*" (p. 36; italics added).

answers, or what? It is not always easy to tell, and even if it sometimes is easy, a theory of answers to why-questions needs to contain a theory of each of these things, and a "theory of explanation"—a theory that only provides a completion of schema (9)—can only be one of them.

I have accused Carl Hempel of being confused, of seeking a theory of explanation when he should have been after a theory of answers to why-questions. Hempel was smart and beloved by many and I do not want to be mean. So I should say that my accusation is not entirely fair, because Hempel did explicitly connect explanation, or at least the kind of explanation he was interested in, to why-questions. He came right out and said "A scientific explanation may be regarded as an answer to a why-question" ("Aspects," p. 334).[21] But Hempel was not willing to characterize his subject matter as answers to why-questions. He denied that just any answer to a why-question constituted a scientific explanation.

Of course Hempel was right. Not all why-questions count as scientific (nor, as I have said, do all acts of answering count as explanations).[22] But that's not relevant here. Even if we ignore non-scientific why-questions, Hempel still limited his attention to only some why-questions, and the question is whether this was a good idea.

Hempel claimed that not all why-questions are "explanation-seeking" why questions; some are instead "epistemic." He took explanation-seeking why-questions to be the target of his theory. Hempel's distinction has been quite influential. Nearly forty years later Salmon reports no challenges to it, and does not challenge it himself, in his book-length history *Four Decades of Scientific Explanation* (see pp. 6 and 136). I don't know of any

---

[21] It is worth noting that after this sentence Hempel writes some things that show he was sensitive to the idea of a noun phrase functioning as a concealed question.

This is as good a place as any to say something about the terms "explanandum" and "explanans," which Hempel goes on to define after this quotation (he introduced them into philosophy in his 1948 paper with Paul Oppenheim, "Studies in the Logic of Explanation," p. 152; philosophers later found a need also for the plural terms "explananda" and "explanantia"). Given what I have said about the word "explanation," you can imagine how I feel about its Latin relatives. I hate them. I can never remember which means what. I also see no need for them, since we can always talk instead of why-questions and their answers. I therefore refuse to use them, and if this book has any impact, I hope it is to discourage their use. (Alas, sometimes these words will appear in this book in quotations from other writers. When they do I will insert, in brackets, replacements for them.)

[22] I myself don't think there is a philosophically interesting distinction between scientific and non-scientific why-questions, and I will usually ignore the restriction to "scientific" explanations when others make it.

challenges in the twenty-five years that have followed.[23] I myself have invoked the distinction.[24]

Does Hempel's distinction even exist? "Not all why-questions call for explanations," he wrote as he introduced it, "[s]ome of them solicit reasons in support of an assertion" ("Aspects," pp. 334–5). This is hard to understand. Hempel seemed to think that offering evidence ("reasons in support"), and explaining, were exclusive activities. But I can do both simultaneously: I can explain what evidence there is for, say, the proposition that the earth is heating up. And how can it be that not all why-questions "call for explanations"? A natural interpretation makes this the claim that explaining why Q is not always the response called for when someone asks why Q. But since "John explained why Q" is roughly equivalent to "John explained the answer to the question why Q," on this interpretation Hempel comes out saying that sometimes, when we ask a why-question, we are not looking for an answer to the question we are asking. That can't be right.

Hempel did not just assert that there are two kinds of why questions, he tried to give examples:

> statements such as 'Hurricane Delia will veer out into the Atlantic'. . . might be met with the question 'Why should this be so?', which seeks to elicit, not an explanation, but evidence or grounds or reasons in support of the given assertion. Questions of this kind will be called *reason-seeking* or *epistemic*. To put them into the form 'Why should it be the case that p?' is misleading; their intent is more adequately conveyed by a phrasing such as 'Why should it be believed that p?' or 'What reasons are there for believing that p?'. (p. 335)

If Hempel meant that no matter what goes in for "Q," "why Q?" has two readings, one explanation-seeking and one epistemic, then he is clearly wrong. If we could say "Why will Hurricane Delia veer out into the Atlantic?" as a way to request evidence, then (given what we know about

---

[23] Though see footnote 25 for a recent challenge to the existence of an epistemic sense of "because."

[24] In "Are There Genuine Physical Explanations of Mathematical Phenomena?" (p. 72, note 2).

the world) we could hear "Because the weatherman said it would" as a good answer. But we cannot.[25]

Instead of asserting that every why-interrogative has an epistemic reading, Hempel may have meant to assert that why-interrogatives with certain forms are epistemic (and can only be read as epistemic), those forms being closely related to the form "Why should it be believed that P?" Hempel is certainly right that there is a set of why-questions that can be asked by uttering a sentence of this form, and a set of why-questions that cannot. But what is the justification for restricting one's theory only to the second set of why-questions? It is silly to reply "the justification is that those are the *explanation-seeking* why-questions." Why call them the explanation-seeking why-questions? Why think that there is a theoretically interesting difference between them and the epistemic why-questions?

True, any bit of evidence you like can appear in an answer to the question why it should be believed that Delia will veer into the Atlantic; not just any bit of evidence can appear in an answer to the question why (in fact) Delia will veer into the Atlantic. This does not show that one cannot have a general, unified theory of answers to why-questions. The difference between the answers could be traced to differences in what follows "why" in each case, not to a difference in the meaning, or use, of the word "why" itself.

After all of this complaining, I am going to agree with the spirit of what Hempel is doing here. I think we should start off restricting our attention to only some why-questions. But my restriction is different from Hempel's.

I'm going to isolate the why-questions I want to start thinking about by looking at the form of their answers. Why-questions admit more kinds of answers than other wh-questions. A "canonical" answer to "Who came to

---

[25] We could hear it as a good answer, of course, if we came to believe that the pronouncements of weathermen could influence the weather.

Several authors claim that "because" is ambiguous. Schnieder, for example, in "A Logic for 'Because',", argues for a distinction between an epistemic and a "genuinely explanatory" meaning of "because." One might worry that, if this distinction is real, then Hempel is right, for he could say that epistemic why-questions are those answerable using the epistemic sense of "because" and explanatory why-questions are those answerable using the explanatory sense of "because." Jonathan Shaheen refutes the claim that "because" has an epistemic sense in chapter 3 of *Meaning and Explanation*.

the party?" can only be expressed (in English) by a sentence of the form
"X came to the party." Contrast this with the following pairs:

- Why did the car skid off the road?
  It skidded off the road because it had a flat tire.

- Why did Jones turn left?
  Jones turned left in order to avoid the cliff.

There are evidently at least two forms that answers to why-questions can
take. They can use "because," or they can use "in order to." "In order to"
answers are connected with ends or purposes: teleology. Because that is a
fraught topic I am going to set aside why-questions that demand "in order
to" answers for later. I will discuss the relationship between "because"
answers and "in order to" answers in section 6.1.[26]

I will also set aside some why-questions that take "because" answers.
In her book *Intention* Elizabeth Anscombe said that an intentional action
is one "to which a certain sense of the word 'Why?' is given application;
the sense is of course that in which the answer, if positive, gives a reason
for acting" (section 5). For now I want to set aside why-questions that are
requests for such reasons. This excludes some why-questions that demand
"in order to" answers, for "in order to" can be used to give someone's
reason for acting—the most natural way to hear "I am walking to the store
in order to buy some milk" is as equivalent to "My reason for walking to
the store is to buy some milk." It also excludes some why-questions that
take "because" answers, for "because" can also be used to give someone's
reason for acting—the most natural way to hear "I am walking to the store
because it sells milk" is as equivalent to "My reason for walking to the
store is that it sells milk." I am making these restrictions, first, because I
want to start with the kinds of why-questions that philosophers of science

---

[26] I should say that using "in order to" is not the only way to answer a why-question by
supplying a purpose. One may also say "Jones turned left with the aim of avoiding the cliff,"
or "Those gutters are there to catch the rain," or "Those gutters are there so that they may
catch the rain." One may also use "for" to supply a purpose: "Why did you wash the dog?—I
washed the dog for fun." And there are yet other constructions for supplying a purpose (see
Huddleston and Pullum, *Cambridge Grammar*, pp. 727–31). For convenience I will call any
answer to a why-question that supplies a purpose an "in order to" answer.

Similarly, answers to why-questions that use "because" can be expressed without the word
"because." More on this in footnote 2 in chapter 3.

have most focused on,[27] and second, because these kinds of questions are more fundamental, in the sense that a theory of answers to them can be used to construct theories of the other kinds of answers. I will lift these restrictions in chapter 6.

2.2 More Red Herring

This section is really a glorified appendix; its aim is to back up my claim that taking the term "theory of explanation" as a serious guide to one's subject matter can lead to irrelevant theorizing.

Peter Achinstein, in *The Nature of Explanation,* held that existing theories of explanation, like for example Hempel's DN model, were inadequate, because they do not say what it takes for someone to explain why Q—to perform the speech act of explaining with respect to the question why Q.[28] Hempel's DN model, he complained, merely attempted to say what it takes for something to be *an explanation* of the fact that P, where "explanation" is used as a name for the product of an act of explaining, rather than the act itself. Achinstein also held that the DN model was false even here; he held that it made false claims about what it took to be the product of an act of explaining.

Suppose Dr Jones explains why Sam died one hour after being operated on by saying: Sam had disease D at the time of his operation; anyone with disease D at the time of an operation dies one hour after the operation; so Sam died one hour after his operation.[29] Suppose that it was Dr Smith who operated on Sam, and Dr Smith offers the very same argument as an excuse for his failure to save Sam. Achinstein argues: if the DN model of explanation is right, then the excuse given by Dr Smith is identical to the explanation (in the product sense) given by Dr Jones. It follows that the explanation given by Dr Jones is an excuse, which is false—Dr Jones had no intention of excusing anything. Explanations (in the product sense) cannot be arguments, Achinstein concludes; instead they are ordered pairs, the first member of which is an argument, the second member of

---

[27] Philosophers of science have certainly focused on "teleological explanation," that is, questions with "in order to" answers, but it has not been a popular topic in recent years.

[28] Achinstein focused more generally on "explaining X," where indirect questions other than why-questions can go in for "X." For simplicity I stick with the case of why-questions.

[29] This is, almost word for word, Achinstein's own example. The example, and his argument concerning it that I am about to discuss, appear on pp. 82–3.

which is an act-type. The explanation Dr Jones gave is the ordered pair, the first member of which is the argument he gave, the second member of which is the act-type of explaining why Sam died one hour after his operation.[30]

What is going on? Hempel said he was offering a theory of explanation. Achinstein sees only two things that a theory of explanation can be: a theory that states necessary and sufficient conditions for an utterance to be a instance of the speech act of explaining, or a theory that states necessary and sufficient conditions for something to be the product of someone's act of explaining.[31] Even if he were right that Hempel's DN model, when interpreted as a theory of the second sort, is false,[32] it would not be much of an objection. The DN model is much more plausible, and more interesting, when interpreted as a theory that states necessary and sufficient conditions for something to be an answer to the question why Q.[33] Maybe the conditions on being an answer to the question why Q are different from the conditions on being someone's explanation of why Q; if so, it is the former that philosophers of science are going to be interested in. By getting hung up on the word "explanation" Achinstein has forced the DN model into the wrong mold.

For another example of how over-use of "explanation" can obscure what is really going on, consider Peter Lipton's essay "Understanding without Explanation." He begins the essay like this:

> Explaining why and understanding why are closely connected. Indeed, it is tempting to identify understanding with having an explanation. Explanations are answers to why questions, and understanding, it seems, is simply having those answers.   (p. 43)

---

[30] This isn't exactly Achinstein's view; it is what his view would be if he were sympathetic to the DN model. His view is that the first member of the ordered pair is a proposition, not an argument. It is a proposition that is an answer to the relevant why-question. See section 3.6 of *The Nature of Explanation*.

[31] Achinstein announces that his aim is to produce theories of these two kinds on p. 3.

[32] He's not. From the proposition that the explanation given by Dr Jones is an excuse it does not follow that Dr Jones intended to excuse himself or someone else for something he or they did. His explanation can be an excuse without him having offered it as an excuse.

[33] On p. 101 Achinstein seems to say that the DN model can be "reformulated" as a theory of this kind. I don't think this is a reformulation of the theory, it is the theory as it should have been understood in the first place.

Lipton, however, wants to resist this identification: he holds that understanding is to be identified, not with having an explanation, but with
"some of the cognitive benefits of an explanation" (p. 44). He goes on to
argue that "there can be understanding without explanation" (p. 45).

But Lipton's use of the word "explanation" as a stand-alone noun makes
his thesis look more interesting than it really is. For what, precisely, is
his thesis? One route to an interpretation goes like this: explanations are
answers to why-questions; so having an explanation is having an answer
to a (relevant) why-question; the relevant kind of understanding is understanding why such-and-such is the case; so "there can be understanding
without explanation" means that someone can understand why Q without
"having" an answer to the question why Q. But this is still unclear: what
does it take to "have" an answer to the question why Q? On one natural
interpretation, one has an answer to the question why Q iff one knows an
answer to the question why Q. Lipton's thesis becomes the claim that one
can understand why Q without knowing why Q. That sounds pretty wild.
Surely knowing why Q is necessary for understanding why Q. Lipton's
thesis, on this first interpretation, is pretty clearly false.

Lipton's examples of "understanding without explanation" suggest a
second interpretation of his thesis. "I never properly understood the why
of retrograde motion," he writes,

> until I saw it demonstrated visually in a planetarium . . . These visual
> devices may convey causal information without recourse to an
> explanation. And people who gain understanding in this way may
> not be left in a position to formulate an explanation that captures
> the same information. Yet their understanding is real.   (p. 45)

The example seems to be one in which Lipton understands why some
planets engage in retrograde motion, without anyone having explained
to him why they do so, and without him being able to explain to anyone
else why they do so. Since the example is meant to support Lipton's thesis,
his use of it suggests that his thesis is that someone can understand why
Q without anyone having explained to her why Q, and without her being
able to explain why Q. But who would deny this?

The second interpretation, while it may fit the example, does not fit
at all with the way Lipton introduced his thesis, as the claim that one
may have understanding without "having" an explanation. Just because
no one explained to him why planets engage in retrograde motion, just

because he cannot explain it to anyone else, does not mean he does not "have" an explanation. In general one can know X (for example, why some planets engage in retrograde motion) without being able to articulate one's knowledge verbally; surely, similarly, one can have an explanation without being able to articulate that explanation verbally (this is certainly true if having an explanation of why Q just is knowing why Q).

I have belabored the point. Let us leave explanation, and "explanation," behind, and turn to our proper topic, answers to why-questions.

3

Reasons Why Are Causes
or Grounds

3.1 From Why-Questions to Reasons Why

A theory of "because" answers to why-questions (that do not give an agent's reason for acting) fills in the right-hand side of the schema

(1) Q because R if and only if . . . [1]

I said at the end of section 2.1 I would give a theory like this, but that was false advertising. I will give a theory, not of answers of the form "Q because R," but of answers of the form "that R is a reason why Q," or "one reason why Q is that R."

It wasn't really false advertising, though, since "reasons why" answers and corresponding "because" answers are often equivalent, in some sense. In response to the question "Why is it raining?" one might answer

(2) It is raining because a cold front has moved in.

Alternatively, one might answer

(3) The reason why it is raining is that a cold front has moved in.

There may certainly be stylistic reasons for preferring (2) to (3). But the *answers one gives* by uttering (2) and by uttering (3) seem to be the same, the two sentences just being different means for giving it. I'm not saying that (2) and (3) are synonymous sentences; if sentence-meanings are fine-grained enough, the fact that (2) and (3) do not have the same surface syntactic structure is good evidence that they are not synonymous.

---

[1] Really a theory of "because" answers should aim for a necessitated biconditional, or something stronger. I will return to this point later in the chapter.

Nevertheless, they certainly seem to express equivalent propositions. Someone who uttered (2) would not add anything to her answer by saying (3), and vice versa.

The fact that (2) and (3) are equivalent suggests a general hypothesis connecting "reason why" answers to "because" answers. In general, there may be many reasons why Q. When there is only one relevant reason why Q, and it is that R, then (and only then) is it the case that Q because R.[2]

Is this right? Some support comes from the fact that it is common in informal speech to begin answers to why-questions with "The reason is because . . . "[3] But this evidence is not very strong. Better evidence might come from contemplating a scenario with more than one relevant reason why something happened. Suppose that Billy distracts the guards while Suzy throws the rock, that the guards would have intercepted the rock had they not been distracted, that you know these things, and that I do not. I ask why the window is broken. In this context there are two relevant reasons why the window broke: that Suzy threw, and that Billy distracted the guards. Would it be wrong for you to answer my question by saying "because Suzy threw a rock at it"?

I don't have clear intuitions. Maybe the hypothesis is wrong. Another possibility is that "because R" is equivalent, not to "that R is the reason" but to "that R is a reason." Or maybe both of these are wrong and "because R" can be true even if that R is not itself a reason, as long as it is a conjunction of reasons.

It does not matter for my project which, if any, of these hypotheses is true. My goal is a theory of "reason why" answers to why-questions; figuring out exactly how they relate to "because" answers can wait.

I am focusing on "reason why" answers for several reasons. One is that doing so lights the way to the best precisification of the vague idea

---

[2] Using "because" or "reason why" is not the only way to report a reason why. One may say also, for example, "Since his train was delayed, he was late for the meeting." I ignore ways of reporting reasons other than using "because" or "reason why" because, first, for my purposes they can all be paraphrased using "because" or "reason why," and second, they cannot be used to answer why-questions. (To "Why was he late?" one cannot answer "Since his train was delayed." See Huddleston and Pullum, *Cambridge Grammar*, pp. 731–32 for this point, and for a catalogue of the linguistic devices usable for reporting reasons.)

[3] Some people think that "The reason is because" is a confused and ungrammatical mash-up of two forms of answer. Huddleston and Pullum deny that it is ungrammatical, holding instead that it is a grammatical construction that is discouraged in formal contexts (*Cambridge Grammar*, p. 731).

that "explanations cite causes." Another is that the notion of a complete answer, and of a partial answer, to a why-question, can be analyzed in terms of the notion of a reason why. I think that these analyses (I will get to them in section 3.3) should be not be controversial; they should be common ground. What *should* be controversial, what philosophers should be arguing about, is what it takes to be a reason why. I will say more about these, and other, virtues of focusing on "reason why" answers over the course of this chapter and the next.

Less important to me, but also worth noting, is that speaking of the reasons why such-and-such happened is a more flexible way of talking. It lets us make more fine-grained distinctions. "Reason," as used in "reason why Q," is a count noun, so we can speak of *a* reason why Q, or *the* reason why Q; we can discuss some or all or just three of the reasons why Q. This flexibility is absent if we are limited to saying things of the form "Q because R."[4]

Again, what matters for my theory is that there are "reason why" answers, not how they relate to "because" answers. And there *are* "reason why" answers, and they are perfectly ordinary. They pervade our lives. "Reason why" is not a special technical term that I am here introducing for the first time. If you doubt this, spend a few days looking, or listening, for "reason why" answers. I never noticed them until I started caring about them, but once I did I found them everywhere.[5] Here are two of many recent examples. I was recently reading "Davidson's Theory of Intention" by Michael Bratman. I selected it more or less at random. Partway through section II Bratman writes, "this cannot be his [Davidson's] considered view. Let me note three *reasons why*" (p. 19; my emphasis). A second example: not long ago I gave a talk about why-questions. Outside the seminar room was a notice that began with something like,[6] "There are

---

[4] Of course, this is not the only grammatical construction containing "because." We can also say "The game was called because of the rain," where "because" is followed by a prepositional phrase rather than a clause. This, however, just seems like a short way of saying that the game was called because it started raining.

There is a way to get some of the flexibility of reasons-why talk using "because," by saying things like "The tree fell because of the following three facts: . . . "

[5] It is like buying a car. Before I bought a Mazda I would have said that there are not many of them on the road. Afterwards I couldn't believe how many I saw.

[6] Unfortunately I did not copy down the exact wording.

two *reasons why* you may not have appeared on the department's most recent list of philosophy majors . . . " (my emphasis).

Some linguists mention reasons when they discuss why-questions. Huddleston and Pullum, in their grammar book's section on questions and their answers, write that "*When, where,* and *why* call for replacements denoting times, places, and reasons, respectively."[7] Or consider Lauri Karttunen's semantics for wh-questions in his seminal paper "Syntax and Semantics of Questions." Roughly speaking, Karttunen says that while declarative sentences express propositions, interrogative sentences "express" (have as their semantic value, for the purposes of compositional semantics) *sets* of propositions. For example, the semantic value of "Who came to the party?" at a world W is the set of propositions P satisfying (i) there is a person X such that P is the proposition that X came to the party, and (ii) P is true at W. Notice that the conjunction of this set of propositions is the true and complete answer to the question that "Who came to the party?" asks in W. Now the semantic value of "Who came to the party?" is a function of the semantic values of its constituents, and so is in part a function of the semantic value of (the interrogative use of) "who." Karttunen holds that "who" has the same semantic value as the existential quantifier "someone." The "there is a person" in clause (i) comes from this meaning of "who." Karttunen does not discuss why-questions, but Jason Stanley suggests that the natural extension of his theory to why-questions treats interrogative "why," like "who," as an existential quantifier. But instead of ranging over people, like "who" does, "why," Stanley suggests, ranges over reasons.[8] "Why" contributes the "there is a reason" to this specification of the semantic value of an interrogative "why Q":

---

[7] *A Student's Introduction*, p. 166; see also *Cambridge Grammar*, p. 725.

When an interrogative word or phrase functions in the clause as something other than a subject (for example, as the direct object of the verb), it usually moves to the front of the clause, as in "Which guitar did Keith Richards play on the Stones' most recent tour?" However, the interrogative phrase sometimes remains in its "original" position, for example when the sentence is being used as an echo question, or in "quizzes or game shows." A game show host might phrase the question above like this: "Keith Richards played which guitar on the Stones' most recent tour?" (see *A Student's Introduction*, p. 165). If we accept that "why" calls for a replacement denoting a reason, then further evidence that "Q because R" entails "that R is a reason why Q" comes from the fact that one non-fronted use of interrogative "why" is in sentences like "John went to the store because why?".

[8] *Know How*, p. 45. Much the same view appeared earlier in Achinstein, *The Nature of Explanation*, p. 30.

- The semantic value of "why Q" at W is the set of propositions X satisfying (i) there is a reason R such that X is the proposition that R is a reason why Q, and (ii) X is true.[9]

I have yet more reasons for focusing on "reason why" answers to why-questions than I have given here. But it is best to present them after I present my theory.

3.2 A Simple(ish) Theory

My theory of reasons why starts out looking fairly simple, but in the end is complex enough that it is best to work up to it in stages. It is a theory of reasons why Q when the fact that Q "corresponds to the occurrence of a concrete event." The fact that a certain firecracker exploded corresponds to a concrete event, namely the explosion. The fact that $2 + 2 = 4$ does not. I will not try to give a precise definition of "concrete event," or much of a theory of which concrete events there are, or a rule for determining which event corresponds to a given fact. It is never going to matter which event corresponds to a given fact; and only some very general claims about which events exist, and are concrete, will be important.

I'm a liberal about what events there are. Events don't have to be changes, like the explosion of a firecracker, or the swerving of an atom. If a room is completely empty for an hour, there is an event that occurs in that room during that hour. I am not, however, an *extreme* liberal about events. I don't think that every fact corresponds to an event; I also doubt that every fact "about some particular things" corresponds to an event. Now if, when considering objections of the form "One reason why Q is that R; but according to your theory this is not a reason," I were going to frequently defend my theory by denying that the fact that Q corresponds to an event, and so falls outside the scope of my theory, I'd need a more detailed theory of events to show that there is something systematic in

---

[9] On another approach to providing a compositional semantics for interrogatives, due to Hamblin in his paper "Questions in Montague English," the semantic value of "who" is not a quantifier but instead the set of all people. Presumably on this approach the semantic value of "why" is the set of all reasons—so it still connects "why" to reasons.

my denials. But I'm not going to defend my theory this way—at least not often, and we can talk about the cases as they arise.[10]

Concrete events include, among other kinds, physical, biological, sociological, geo-political, and mental events.[11] So my theory makes claims, not just about the reasons why a given rock accelerated when dropped but also about the reasons why the economy collapsed in 2008, and even the reasons why a given non-physical mind (if there are any such things) thought "I think." What my use of "concrete event" excludes are "events" concerning, for example, only mathematical objects, like— to use my earlier example—the event consisting in the numbers 2 and 4 being related such that $2 + 2 = 4$. I imagine most philosophers would deny that there is any such event; even if they're wrong, my theory says nothing about the reasons why $2 + 2 = 4$.

A first draft of my theory consists of instances of the following schema, in which "Q" is replaced by a sentence expressing a fact that corresponds to the occurrence of a concrete event:

(T0)   That R is a reason why Q if and only if the fact that R is a cause of the fact that Q.

Many philosophers have believed that "explanations of events cite, or describe, causes." They do not all construe this vague claim in the same way. Theory (T0) is, I think, the best way to make it precise. I will say something about the ways in which it is superior to another well-known precisification, David Lewis's, below.

(T0) has a lot going for it. The car skidded off the road. One reason why is that the tire blew out. The fact that the tire blew out is also a cause. Again, the main reason why the dinosaurs went extinct is that a comet (or asteroid) hit the earth. The comet impact also caused the extinction. And again, one reason why there was so much snow in Boston last winter is

---

[10] Maybe I should also say that I am a "superficialist" about events. I'm willing to talk about them, even to use event-talk when doing philosophy (as is obvious by now), but I don't really believe there are such things as events. If this were a book on ontology I would avoid event-talk, or show how it may in principle be avoided. This is not a book on ontology so I will indulge in event-talk, since it makes many things easier to say. The event-talk is there just to delimit which facts my theory of reasons why applies to.

[11] Of course on some views every biological, sociological, and so on event is a physical event. But this has been disputed.

that humans have released too much carbon dioxide into the atmosphere. This was also a cause of the snow.

I could go on and on. Examples that corroborate (T0) are easy to find.

Nevertheless, I have to acknowledge that (T0) is false, in fact quite obviously false. For example, one reason why this ball is either red or green is that it is red. But the fact that the ball is red is not a cause of the fact that it is either red or green.

If this reason why is not a cause, what kind of thing is it? It is a ground. The fact that the ball is red grounds the fact that the ball is either red or green. In general, "grounding explanations" of events are counterexamples to (T0).

The example of the ball certainly involves a reason that is a ground, but is that reason really a reason why some event occurred? Is there really such an event as the ball's being either red or green? I'm liberal enough to say yes, but this question can, as Judith Thomson likes to say, be bypassed. Even if there is no such event as this ball's being either red or green, there are other examples of reasons why that are grounds that are uncontroversially reasons why some event occurred. The temperature of the air in this room just increased from 71 degrees Fahrenheit to 72 degrees Fahrenheit. Why? Because the average kinetic energy of the molecules that constitute the air in this room just increased from 6.10×10^{-21} Joules to 6.11×10^{-21} Joules.[12] The increase in kinetic energy did not cause the increase in temperature; it is the ground for the increase in temperature.

So there exist reasons that are grounds as well as reasons that are causes. But that is where I draw the line. Here is a preliminary statement of my theory:

(T1p) That R is a reason why Q if and only if the fact that R is a cause of the fact that Q, or a (partial[13]) ground of the fact that Q.

Carl Hempel entertained, but then rejected, the idea that "all explanations of events are causal." The main example he gave to motivate this rejection is the following answer to the question why the period of some given pendulum is two seconds: this pendulum has a period of two seconds

---

[12] Assuming that the air in the room is an ideal gas.

[13] There are analogies between the structure of causation and the structure of grounding, and in Kit Fine's terminological scheme it is "partial ground" that corresponds to "cause" ("Guide to Ground," p. 50). Having said this, I will often use "ground" to mean "partial ground."

because it is one hundred centimeters long. But "surely" the length does not cause the period ("Aspects," p. 352). Now "this pendulum has a period of two seconds because it is one hundred centimeters long" is true only if we interpret the talk of the pendulum's period as talk of a disposition. Phrased in terms of reasons why the claim then becomes: one reason why the pendulum is disposed to take two seconds to complete a full swing is that it is one hundred centimeters long. Hempel is right, the length does not cause the dispositional period. Once again we have a counterexample to (T0), but once again it is compatible with (T1p). While the length may not cause the period, it does ground it.

David Lewis entertained, and then accepted, the idea that "all explanations of events are causal" (in his paper "Causal Explanation"; I will have a lot more to say about this paper in the next section). Reasons that are grounds refute Lewis's theory. "Ground" was not in the official philosophical lexicon when Lewis wrote his paper, but he does discuss an example that we would today characterize in terms of ground, and try to show that it is compatible with his theory:

> Walt is immune to smallpox. Why? Because he possesses antibodies capable of killing off any smallpox virus that might come along. But his possession of antibodies doesn't *cause* his immunity. It *is* his immunity. Immunity is a disposition, to have a disposition is to have something or other that occupies a certain causal role, and in Walt's case what occupies the role is his possession of antibodies.
>
>   I reply that it's as if we'd said it this way: Walt has some property that protects him from smallpox. Why? Because he possesses antibodies, and possession of antibodies is a property that protects him from smallpox. Schematically: Why is it that something is F? Because A is F. An existential quantification is explained by providing an instance. I agree that something has been explained, and not by providing information about its causal history. But I don't agree that any particular event has been non-causally explained. The case is outside the scope of my thesis. That which protects Walt—namely, his possession of antibodies—is indeed a particular event.... But that was not the explanandum [the thing being explained].... What we did explain was something else: the fact that something or other protects Walt. The obtaining of this existential fact is not an event. It cannot be caused.   (p. 223)

REASONS WHY ARE CAUSES OR GROUNDS   31

Lewis had a quite detailed theory of events (presented in his paper "Events"), and he goes on to appeal to it to justify his claim that there is no such event as Walt's immunity. I don't accept Lewis's theory of events, so I don't accept Lewis's defense of his theory. Nor does what Lewis say in this passage seem plausible, evaluated on its own. Walt's immunity to smallpox cannot be caused? Go to the press and announce that philosophers have discovered that no one at the Centers for Disease Control has ever immunized anyone. We'll be laughed out of town. Surely injecting someone with a smallpox vaccine—something plenty of CDC agents have done—is a way to immunize that person against smallpox; and certainly whenever someone immunizes someone against smallpox, he causes that person to be immune to smallpox. Walt's immunity *can* be caused; so it *is* an event.

I think it is futile to try to defend (T0) by taking on some weird theory of events. Not only will the theory need to say that there is no such event as Walt's immunity to smallpox, it will also need to say that there is no such event as the air in the room's having increased from 71 to 72 degrees Fahrenheit. For if there is such an event, then the truth of "the reason why the temperature increased is that the mean molecular kinetic energy increased" will be a counterexample to the theory. But surely *that* event (if it exists!) can be caused. Didn't I just cause it by turning up the thermostat?

What we should do is abandon (T0) and move to (T1p). Walt's immunity is grounded in his possession of antibodies. That's why his having the antibodies is a reason why he is immune.

Sometimes, when someone asks why the temperature increased, we answer that the thermostat has been turned up. Other times, we answer that the mean molecular kinetic energy has increased. Theory (T1p) says that these are both correct answers. Still, it might seem bad to give both of them at once. "The temperature increased for several reasons, among them: the fact that someone turned up the thermostat, and the fact that the mean molecular kinetic energy went up." I don't find this particularly odd, but I imagine someone might. Even if it is odd, and my judgment that it is fine is off, I don't see a threat to (T1p). Maybe the right thing to say is that in some contexts we use "why" to ask for reasons why that are causes, and in other contexts we use "why" to ask for reasons why that are grounds, and we never use "why" to ask for both kinds of reasons simultaneously. It's not crazy to think that we make this restriction, since causation and grounding are different things. All of this is compatible with the claim that, speaking unrestrictedly, causes and grounds, and only causes or grounds, are reasons why.[14] Another reaction, defended by Jonathan Shaheen (*Meaning and Explanation,* chapter 2), is that "reason" in "reason why" is ambiguous (or, since the meanings are closely related, "polysemous") between a causal meaning and a grounding meaning. My theory then would be that in one sense of "reason," all and only causes are reasons why, and in the other sense, all and only grounds are reasons why.[15]

Theory (T1p) raises the question: what is grounding? I do not have a lot to say about this question. I have given examples: the temperature of a gas is grounded in the average kinetic energy of its constituent molecules, a disjunction is grounded in its true disjunct(s). If these sound right then you already sort of know what grounding is. Another way into the notion of grounding is to trace its connections to other notions. For example, fact G grounds fact H iff G is a metaphysically more basic fact in virtue of which H obtains. That is not much help since "H obtains in virtue of the fact that G obtains" is close to synonymous with "H is grounded in G." But few of those who theorize about grounding give reductive definitions of grounding. They do not try to say in more basic terms what grounding is. I am convinced that the notion of ground is intelligible, and will make use of it without apology or further explanation.[16]

To the extent that (T1p) raises the question of what grounding is, it also raises the question of what causation is. I do not have a lot to say about this question either. In contrast with ground, reductive theories of causation are popular. But I am not going to defend a detailed theory of causation in this book. I do not think that progress in the philosophy of reasons why must wait for a complete theory of causation (good thing, or we will wait

---

[14] The view I am floating here is that in different contexts we restrict the domain of quantification to different kinds of reasons (recall from earlier in this section that the Karttunen semantics for "why" treats it as a quantifier over reasons). One natural alternative to this view is to say that "reason" in "reason why" is itself context-sensitive. I will say something about this view at the end of section 3.4.

[15] While I am not convinced that "reason" is ambiguous between a causal and a grounding meaning, I do think it is ambiguous; one meaning is used when giving an agent's reasons for acting, the other is not. Since reasons for action are off the radar right now, I will not defend this claim until Appendix D, in chapter 6.

[16] An extended defense of the intelligibility of the notion of ground may be found in Gideon Rosen's paper "Metaphysical Dependence: Grounding and Reduction."

forever), nor do I think that a theory of reasons why needs to be combined with a detailed theory of causation for us to be able to evaluate it.

It seems that there was a time, during the fevered heyday of logical positivism, when it was thought illegitimate to use the notion of causation in one's theory of explanation, if one did not have a worked-out reductive theory of causation. Maybe the idea was something like this: "surely causation is at least as problematic, probably more problematic, than the notion of explanation; so analyzing explanation in terms of causation is no progress at all."[17] I am glad those days are over. Showing how one notion is connected to another can be philosophically illuminating, even if both notions are "philosophically problematic." Finding that connection can be a step towards reducing all philosophically problematic notions to unproblematic notions.[18] When one finally works out one's complete theory of everything one will of course want a theory of causation, but I am young.

Maybe you don't think there can be a "theory of everything," if that's a theory that reduces all problematic notions to unproblematic ones. Maybe you don't accept a division of notions into the problematic and the unproblematic. You should (or, could) still think it a good thing to have a theory of reasons why. Finding a good theory can constitute philosophical progress even if you reject this division.

In theory (T1p) causation is treated as a relation between facts rather than events. I prefer "the fact that Suzy threw a rock at the window was a cause of the fact that the window broke" to "the throw was a cause of the breaking"—at least when it comes to choosing a way of talking to build a theory of causation around. But I'm not going to defend here the thesis that a theory of causation should treat causation as fundamentally a relation between facts.

In fact—though this is a bit of a side-note—I am sympathetic to the idea that causation is not a relation between facts either. I like the idea that

---

[17] Jaegwon Kim, in his paper "Hempel, Explanation, Metaphysics" (p. 11), speculated that Hempel would have held this view. (Philip Kitcher has assured me in conversation that this was definitely Hempel's view.) Positivism's heyday preceded World War II, yet still in 1976 we find Larry Wright observing "an unhappy tendency in the literature of analytical philosophy to feel that the unanalyzed notion of a causal connection is logically so septic that it should always be avoided in favor of some—almost *any*—reduction formula" (*Teleological Explanations*, p. 32).

[18] Of course what counts as problematic has varied over the course of philosophy.

"causation is not a relation" in the first place. That is, I like the idea that the proper way to regiment causal talk, for philosophical purposes, is to use, not a predicate "X is a cause of Y" that is true of two things only if both are facts (or both are events), but to use a sentential connective that takes two sentences and makes a sentence. If we use "C" for this connective, then "Suzy's throw caused the breaking" and "The fact that Suzy threw a rock is a cause of the fact that the window broke" get replaced by "C(Suzy threw a rock; the window broke)." There seem to be expressions in English the function of which is close to that of "C": for example, "Suzy threw a rock, *and as a result*, the window broke." In fact, I think that "The fact that . . . is a cause of the fact that . . . " is another such expression. (I am saying that these expressions have similar functions, not that they are synonymous; they have syntactic, and therefore presumably semantic, complexity missing from "C(. . . ; . . .).")[19]

One objection to treating causation as a relation between facts, rather than events, is that it makes causation too "unworldly": surely it is things in space and time that cause other things in space and time, but it is events that happen in space and time; facts are "out there" with the other abstract objects, the numbers for example, and numbers are the wrong kind of thing to do any causing. The thesis that causal talk is best regimented using a sentential connective, while in many respects very similar to the thesis that causation relates facts, avoids this objection. If asked to identify the thing doing the causing, and the thing being affected, in the sentence "Suzy threw a rock, and as a result, the window broke," the best answer is that it is Suzy and the rock (both worldly things in time and space) that do the causing, and the window that is affected.

Anyway, (T1p) can be reformulated to treat causation as a relation between events. It will say: that R is a reason why Q iff the fact that R is a ground of the fact that Q, or the event corresponding to the fact that R is a cause of the event corresponding to the fact that Q.[20]

---

[19] Van Inwagen, in "Causation and the Mental," asserts this theory of causation, suggests the translation of "C" as "and as a result," and also suggests other translations.

[20] (T1p), reformulated in this way, is most plausible if we assume a plenitudinous theory of events, one that, roughly speaking, denies that distinct facts ever correspond to the same event. If one held, instead, that, for example, the fact that Jones walked home, and the fact that Jones walked home slowly, correspond to the same event, an event that is both a walking home by Jones and a slow walking home by Jones, then there is trouble; I'll be stuck saying

Having said all this, for stylistic purposes I will sometimes speak of causation as a relation between events, and also sometimes speak as if it is events that are reasons why.

While I am discussing what I am taking causes to be, I should say something about what I am taking reasons to be. It is probably clear that I am taking reasons to be facts (even if, as I just said, I will sometimes write as if reasons are events). It is worth emphasizing, however, that officially, grammatically, the "X" in "X is a reason why Q" and in "one reason why Q is X" holds the place, not for a term denoting a fact, but for a that-clause that expresses a fact. The officially-true thing to say is: "that the car had a flat tire is a reason why it skidded off the road," not "the fact that the car had a flat tire is a reason why it skidded off the road." This is not super-important, and I will sometimes put terms for facts in for "X" when I think it aids comprehension. The difference is still worth noting, though, because it makes my theory immune to a certain kind of objection. If you could put a term for a fact in for "X," then you could answer the question why the car skidded off the road by saying "Fred's favorite fact is a reason why the car skidded off the road," or (worse) "Xvvb is a reason why the car skidded off the road," where "Xvvb" is a name one has introduced for the fact that the car had a flat tire. But these do not seem to be answers. (Saying that Fred's favorite fact is a reason seems instead to be a way to inform someone of how to find out the answer—ask Fred what his favorite fact is.[21])

Jaegwon Kim raised something like this objection to the theory that to answer the question why E happened it is enough to cite causes of E ("Hempel, Explanation, Metaphysics," pp. 16–17). Kim complained that listing some causes of E by name, using some arbitrary system one has

that, if one reason why Jones arrived late is that he walked home slowly, then also one reason why Jones arrived late is that he walked home, even if it is true that had he walked home without walking home slowly, he wouldn't have been late. (As is well-known, Davidson had a "non-plenitudinous" theory of events; see for example his paper "Causal Relations." I guess I should mention that Davidson also objected to speaking of "the" event that corresponds to a given fact. I'm not going to dwell on this, though, since in my official theory it never matters whether a fact corresponds to a single unique event.)

[21] Using terminology I will introduce in the next section, my suggestion here is that "Fred's favorite fact is a reason why the car skidded off the road" is merely a partial answer to the question why the car skidded off the road, while "that it had a flat tire is a reason why the car skidded off the road" appears as a conjunct in the canonical statement of the complete answer to the question why the car skidded off the road.

introduced for naming events (he's assuming that causes are events), is not a way to answer the question. He thought this observation favored some modified version of the DN model, some theory according to which to answer the question why E happened you must exhibit a law connecting the causes of E to it. You cannot construct a DN argument for the conclusion that the car skidded off the road using the premise that Charlie occurred, where "Charlie" is a name for the tire blowout. You will need a premise that describes Charlie a particular way, namely as a tire blowout; for any law connecting flat tires and cars skidding off roads will describe the tire blowouts as tire blowouts.[22] Again, though, Kim's argument does not show something like the DN model to be better than my theory, because my theory does not allow names of, or other terms denoting, facts to appear in a reasons-why answer to a why-question. What must appear is a that-clause that gives the "content" of the relevant fact.

One choice point in the theory of causation is between facts and events; another is between treating the causal relation as a two-place relation and treating it as a four-place relation. "Contrastivists" about causation think it is a four-place relation: a relation between a cause, a contrasting fact/event (or set of contrasts), an effect, and another contrasting fact/event (or set of contrasts).[23] Although I will speak exclusively in binary terms, this is just for convenience; I do not take a stand in this book on whether contrastivism is correct. (If contrastivism about causation turns out to be correct, then I will advocate contrastivism about reasons why as well. On that combination of views, the number of relata of the causal relation will continue to line up with the number of relata of the reasons-why relation.[24])

---

[22] Of course we're pretending here that there are such laws; there are not. The idea that causes and effects are related to each other by laws only under certain descriptions is prominent in Davidson's thinking, for example in "Causal Relations."

[23] See Hitchcock, "Farewell to Binary Causation," and Schaffer, "Contrastive Causation."

[24] Hitchcock simultaneously advocated contrastivism about causation and explanation, in "The Role of Contrast in Causal and Explanatory Claims." Van Fraassen, in *The Scientific Image,* and Garfinkel, in *Forms of Explanation,* were early advocates of contrastive theories of explanation. The contrastivist thesis (about reasons), again, is that the reason-why relation has more than two argument-places. One main motivation for this thesis is that why-interrogatives are often explicitly contrastive. The question why it is raining rather than snowing is different from the question why it is raining rather than not raining. But the contrastivist thesis is not *identical* to the thesis that why-interrogatives are often explicitly contrastive, or to the thesis that they are always at least implicitly contrastive. An argument is needed to get from either of these two theses to the contrastivist thesis, and I am not yet

Theory (T1p), again, applies only when the sentence that goes in for "Q" describes the occurrence of a concrete event. But lots of why-questions are not about events: why is Galileo's law of free-fall true? Why does every complex polynomial have a root? Presumably there is some entirely general theory of reasons why, from which it will follow that, for example, certain facts about gravity are among the reasons why Galileo's law is true. And presumably that entirely general theory will have (T1p) as a special case. What is that general theory?

Oh, how I wish I had a completely general theory of reasons why to show you. But I do not. I have part of a general theory: I certainly think that, in general, if G grounds H, then G is a reason why H obtains. Why not go all the way? Why not take (T1p) and remove the restriction on what goes in for "Q," to get the theory that the schema is true no matter what goes in for "Q" and "R":

(T2)   That R is a reason why Q if and only if the fact that R is a cause of the fact that Q, or a ground of the fact that Q.

When the sentence that goes in for "Q" is true but does not describe the occurrence of a concrete event, then the fact that Q is not the right sort of thing to have causes. (T2) says that, in this case, the reasons why Q are all and only the grounds of Q.

I do not endorse (T2) because I doubt that, in this kind of case, the only way a fact can get to be a reason why Q is by grounding the fact that Q. One general reason for doubting this comes from looking at answers to why-questions about mathematical facts. There are mathematical facts F with the property that the question why F obtains has an answer. But I doubt that every answer to a question like this cites facts that ground that mathematical fact. (I will give another reason for rejecting (T2) in chapter 4, and float a completely general theory of reasons why, one that does better than (T2) but which I still hesitate to fully endorse, in Appendix B, in chapter 4.)

Theory (T1p) connects reasons why to causes and grounds using the biconditional "if and only if." It is common to state theories this way, but read literally the theory is too weak. I don't just hold that, as it turned out, reasons why Q are always either causes or grounds of the fact that Q.

convinced by the arguments that have been given. But, again, taking a position on the debate over contrastivism is not one of my aims in this book.

I hold that this is necessary. More precisely, every instance of the following is true, when "Q" is replaced by a sentence that purports to describe the occurence of an event, and "R" is replaced by any sentence:

- Necessarily, if it is a fact that Q and it is a fact that R, then that R is a reason why Q if and only if the fact that R is a cause of, or a ground of, the fact that Q.

But there is another way in which (T1p) is too weak, that adding "necessarily" does not address. Suppose that it is a fact that Q, and a fact that R, and that the fact that R is a cause of the fact that Q. According to (T1p), that R is a reason why Q. This is a book about why-questions. It is thus natural in this context to ask, about this kind of example, *why* it is that one reason why Q is that R. Although my restriction to why-questions about events entitles me to remain silent about this question, it goes against the spirit of (T1p) to say nothing. It is, instead, in the spirit of (T1p) to answer that it is not just a *coincidence* that (i) the fact that R is a cause of the fact that Q, and also that (ii) the fact that R is a reason why Q. It is in the spirit of (T1p) to assert further that the fact that R is a reason why Q *because* the fact that R is a cause of the fact that Q. Put into reasons-why talk the assertion is that the reason why ⟨one reason why Q is that R⟩[25] is that the fact that R is a cause of the fact that Q.

This iterated reasons-why claim is in the spirit of my theory but does not follow from (T1p). I want it to follow. So a closer-to-official statement of my theory is this:

(T1)   Necessarily, if it is a fact that R and it is a fact that Q, then: if the fact that R is a cause of the fact that Q, then one reason why Q is that R, *and* the reason why ⟨one reason why Q is that R⟩ is that the fact that R is a cause of the fact that Q. Similarly if the fact that R is a ground of the fact that Q. And every reason why Q is (i) either a cause of the fact that Q or a ground the fact that Q, and (ii) satisfies the relevant one of these conditionals.

The final, official statement of my theory is even more complicated than this. But we will not need those complications for a long time, not until

[25] The angles are here to make this sentence easier to parse. They serve this function in the rest of the book.

chapter 5, where I will have a lot more to say about dizzyingly-iterated claims about reasons-why, so I will postpone them.

I already said a little bit about grounding, but putting (T1) up on display is an opportunity to say one more thing. I have treated grounding as a relation between facts. But my sympathy for the idea that causation is not a relation between facts extends to grounding. Maybe the best way to regiment grounding talk, like the best way to regiment causal talk, is to replace a two-place predicate of facts with a two-place sentential connective. This is an idea that already has traction: Kit Fine advocates it ("Guide to Ground," p. 46). But I think it is a mistake to suggest, as Fine does, that " . . . because . . . " is a close English equivalent of this two-place sentential connective. For if "the fact that the ball is red grounds the fact that the ball is either red or green" means pretty much the same as "the ball is red or green because it is red," then it cannot be right to say that[26]

(4)   ⟨the ball is red or green because it is red⟩ because ⟨the fact that the ball is red grounds the fact that it is either red or green⟩.

For if "because" is just a sentential-connective way of expressing grounding claims, then sentence (4) has the form "P because P," and so cannot be true. But (4) *is* true. Although every fact that is a ground is also a reason why, it is false that *being a ground* and *being a reason why* (or, better, that *being a cause or a ground* and *being a reason why*) are the same thing.[27]

3.3 Partial and Complete Answers

"Explanations of events cite, or describe, causes"—many philosophers have accepted this sentence, without agreeing on what it means. Theory (T0) is, I have said, the best way to make it precise. So what is wrong with the other ways? The answer will emerge from looking closely at one alternative precisification.

---

[26] I'm putting it here using "because" rather than "reason why" because that is the terminology Fine uses.

[27] Schaffer makes a similar point in "Grounding in the Image of Causation"—though Schaffer reaches further, trying to argue that neither grounding nor causation should be regimented using a sentential connective. This requires extra premises; holding that (4) is true does not get you that far.

David Lewis held, in his paper "Causal Explanation," that "to explain an event is to provide information about its causal history" (p. 217). Although this is his official statement of his theory (the sentence just quoted follows the declaration "Here is my main thesis:"), throughout his paper Lewis prefers to frame his theory as a theory of what "explanatory information" is. As a theory of explanatory information, the theory says that "a chunk of explanatory information ... is a proposition about the causal history of the explanandum event [event being explained]" (p. 218).[28] Even though Lewis formulated his theory using, first, "explain," and later "explanatory information," I think that Lewis's theory is best interpreted as a theory of answers to why-questions.[29]

As a theory of answers to why-questions, Lewis's theory, like (T0) and (T1), is limited to why-questions about events. The most direct way to translate Lewis's statement of his theory into a theory of answers to why-questions about events gives us this: X is an answer to the question why E happened iff X is a proposition about E's causal history. But putting the theory this way leaves it unclear. There are, as we are about to see, several different kinds of answers to why-questions. Just what kind of answer to a why-question is this theory a theory of?

It is certainly not a theory of reasons-why answers. Lewis never uses the term "reason why." Neither does it appear to be a theory of because-answers. Instead it is most naturally understood to be a theory of partial answers to why-questions.[30]

---

[28] In the part of this quotation that I elided, Lewis says that an explanation, in the product sense, is a chunk of explanatory information. So Lewis also accepted the following as a statement of his thesis: an explanation of E is a proposition about the causal history of E.

Lewis's official statement is of a theory of what it takes to "explain." So officially Lewis proposed a theory of the speech act of explaining. But Lewis was not really interested in what it takes to successfully perform this speech act (see p. 218 for evidence). That's why, I think, he preferred in the rest of the paper to think of his theory as a theory of explanatory information. (As a theory of the speech act, Lewis's theory isn't very plausible. Aren't there many ways to "provide" the answer to a question, not all of which amount to *explaining* that answer?)

[29] Lewis initially mentions why-questions only in passing, by does so to indicate that he sees a tight connection between why-questions and explanatory information: "The why-question concerning a particular event is a request for explanatory information" (p. 218).

[30] Lewis explicitly distinguishes between complete and partial answers to questions on p. 229, but never states his theory as a theory of partial answers to why-questions. Still, this is the interpretation of his theory that fits best with his discussion of contrastive why-questions in section VI, and his critique of the DN model of explanation in section VII. I will discuss part of this critique below.

The distinction between complete and partial answers applies to any question. The complete answer to the question who came to the party says, of all the people who came to the party, that they came to the party (and perhaps add that no one else came to the party).[31] As for partial answers, in general a proposition is a partial answer to a question if it "rules out" (entails the falsity of) at least one possibly-true-but-actually-false complete answer to that question.[32] So "Two people whose names begin with 'J', " if true, is a partial answer to the question who came to the party, since it entails that "Aaron and Moses and no one else came to the party" is not the complete answer.

I take Lewis's view to be that the complete answer to the question why E happened lists all of E's causes (and perhaps adds that they are all of the causes). The part of his theory that he is most explicit about, his theory of partial answers, is what you get by taking the general characterization of partial answers to questions that I just gave, and applying it to this case. Since, in general, a partial answer is something that rules out a possible complete answer, a partial answer to the question why E happened is a proposition that rules out some possible specification of the causes of E.[33]

---

[31] This is an example, not a definition of "complete answer," but I trust it makes the idea clear enough. (Giving a definition requires going in to complications that are not relevant here, such as the distinction between *de re* and *de dicto* readings of questions—for which see Groenendijk and Stokhof, "Semantic Analysis of Wh-Complements.")

[32] This is (almost) the definition of a partial answer standardly-accepted by linguistics working on the semantics of questions—see, for example, Lahiri, *Questions and Answers in Embedded Contexts*, p. 10, where a "partial answer" is defined to be a proposition, which may be false, that rules out (what I would call) a possible complete answer.

It seems to me me that there is, however, another, weaker, characterization of a partial answer: a proposition that reduces the audience's uncertainty about what the complete answer is. A proposition could do this by lowering the probability of a possible partial answer, without entailing that it is false. Having said this, in the text I will stick to the more restrictive, but more common and easier to state, condition.

[33] As we saw, Lewis preferred to speak of "causal histories"; his view is that partial answers give information about E's causal history. But the causal history of E, for Lewis, just is the list of all of E's causes. So there is no benefit to using the term "causal history." But there is a burden. Many philosophers now deny that causation is transitive (see Hitchcock, "Intransitivity," for one example). C may be a cause of D, and D a cause of E, without C being a cause of E. If this is possible, then C is part of the causal history of E without being a cause of E. If causation is not transitive then "the complete answer specifies E's complete causal history" and "the complete answer specifies all of E's causes" are not equivalent. And the first is false: if C is not a cause of E, then (I think) it is not part of the answer to the question why E happened, even if it is part of E's causal history.

I accept these claims... for the most part. I hold that the complete answer to the question why E happened will list all of the *reasons why* E happened (and perhaps add that they are all of the reasons why E happened). A canonical statement of the complete answer will have the form "one reason why E happened is that A, another reason why E happened is that B, another reason why E happened is that C, . . . , and these are all the reasons why E happened." Since I think that every cause is a reason why, I agree that the complete answer specifies all the causes. But I do not accept that the complete answer specifies only causes. I hold that grounds of E are also reasons why E happened.

Like Lewis, I accept that a partial answer to the question why E happened is a proposition that rules out a possible complete answer. But since I disagree with him about what a complete answer looks like, I disagree with him about which propositions are partial answers.

These places where I disagree with (my reconstruction of) Lewis's theory, however, are not where I locate the theory's biggest flaw. Its biggest flaw is that it is incomplete. There is more to a theory of answers to why-questions than a theory of complete answers and a theory of partial answers. A theory of answers to why-questions needs to employ the notion of a reason why, and it needs to contain a theory of reasons why.

A theory needs to employ the notion of a reason why in order to give a general, neutral characterization of complete answers to why-questions. The neutral characterization is (again) this: a complete answer to the question why Q lists all of the reasons why Q. Everyone—those who think that "explanations of events must cite causes," and those who do not—should agree on this. The place for disagreement is over what it takes to be a reason why. We recover Lewis's theory of complete answers to why-questions if we combine this neutral characterization of complete answers with (T0), the claim that the (only) reasons why E happened are its causes. If you think that Lewis's theory is wrong then *this* claim is the one you should put pressure on. Presenting the theory as a theory of complete and partial answers and nothing else makes the target its opponents should aim at invisible.[34]

With reasons why invisible, Lewis's theory can seem "too easy" to defend. Partial answers to why-questions are cheap. Almost any

---

[34] Come to think of it, this is not such a bad strategy.

proposition rules out some hypothesis about what caused E. Why did the window break? The proposition that Jones was at school entails (or, entails given shared background knowledge that the window is far from the school) that nothing Jones did caused the window to break. The proposition that Jones was at school is therefore (in this context) a partial answer to the question why the window broke. But if I ask why the window broke I would not be very satisfied to hear "because Jones was at school." We do not seem to use "because" when we are merely giving a partial answer.[35]

The fact that it is very easy to be a partial answer to the question why E happened has led some philosophers to think that Lewis's theory, or a theory like it, must be wrong. Elliott Sober claimed that a view like Lewis's "trivializes" the notion of causal explanation, in the sense that it makes it "trivially true" that all explanation is causal explanation.[36] Marc Lange, in the course of arguing that a certain explanation is not a causal explanation, admits that the explanation "supplies information about the world's network of causal relations," but claims that this cannot be enough to make it a causal explanation, since "any fact" does that.[37] James Woodward classifies Lewis's theory as one that "appeals to hidden structure," and complains that appealing to hidden structure "makes it too easy to protect one's favored theory of explanation from genuine counterexamples."[38] This is the same objection again: the hidden structure is the complete answer (it is "hidden" because people can exchange explanatory information, in Lewis's sense, without knowing much about the complete answer); Lewis, when presented with explanations that don't identify any causes of the event being explained, maintained that those explanations *do* convey some information, even if only a little bit, about the complete answer (about the causes of the event being explained); given how easy it is for a proposition to do this, Woodward thinks that this strategy for dealing with apparent counterexamples is, somehow, cheating.

---

[35] Of course it's fine to say "I don't know much about why the window broke, but I know that it wasn't because of anything Jones did." Also, with special background information "because Jones was at school" sounds okay—suppose for example that Jones spends all his free time protecting the window. I'm assuming that nothing like this is true.

[36] "Equilibrium Explanation," p. 202; *The Nature of Selection*, p. 141.

[37] "What Makes a Scientific Explanation Distinctively Mathematical?," p. 496.

[38] *Making Things Happen*, p. 175.

This line of opposition to Lewis's theory is aided and abetted by phrasing his theory of partial answers to why-questions as a theory of explanation, as a theory that fills in the right-hand side of

(5)   Fact X explains fact Y if and only if . . .

It certainly sounds wrong to say that the fact that Jones was at school explains the fact that the window broke. So if one thinks that Lewis's theory of partial answers is an attempt to complete schema (5), one will think that it is false.[39]

But the example of Jones and the window does not show Lewis's theory of partial answers, or his theory of complete answers, to why-questions to be false. The example *does* show that it was a public-relations mistake for Lewis to use "explanatory information" as a name for the information contained in a partial answer. Once again, blinkered focus on what is or is not an "explanation" can blind you to what is really going on. There is also another, related, and far more important, moral to draw from the example. The moral is that there is something missing from Lewis's theory. What is missing is a theory of reasons why. The proposition that Jones was at school may be a partial answer to the question why the window broke, but (if (T0) is true) that Jones was at school is not a *reason why* the window broke. That's how to reconcile the thought that this proposition does say *something* (very weak) about the causes of the breaking with the thought that it does not seem to explain the breaking.

The fact that it is very easy for a proposition to be a partial answer to the question why E happened does not make (T0) and (T1) easy (or "too easy") to defend. It just shows that trying to refute these theories by producing a proposition that (i) is in fact a partial answer to the question why E happened, but (ii) is not a partial answer according to (T0) or (T1), is a bad strategy.[40] The *good* strategy is to produce a (true)

---

[39]   When Woodward offers counterexamples to theories like Lewis's (his actual target is Railton), it is in these terms. He argues that there are lots of propositions about the complete, or ideal, explanation of an event that are not themselves "explanatory"; see *Making Things Happen*, pp. 176–9.

[40]   Technically, as stated above, (T0) and (T1) say nothing about what it takes to be a partial answer, or a complete answer. I mean to include in the theories (T0) and (T1) not just the conditions on being a reason why listed in my statements of them above, but also the neutral characterizations of complete and partial answers that I have given in this section.

proposition that (i) is a *reason why* E happened, but (ii) is not a reason why E happened according to one of these theories. For while it is easy for a proposition to be a partial answer to the question why E happened, it is quite hard for a proposition to be a reason why E happened. If one successfully executes this strategy, the proposition one produces will probably be, by the lights of (T0) and (T1), a partial answer to the question why E happened. *It doesn't matter.* If the proposition is not, according to those theories, a reason why E happened, those theories still stand refuted.

Lewis cannot distinguish between these two strategies, and as a consequence his response to the "stellar collapse" example is inadequate. Peter Railton offered Lewis the stellar collapse example as a counterexample to his theory. A star is collapsing, then stops. Why? Lewis accepted the answer "Because it's gone as far as it can go . . . It's not that anything caused it to stop" (p. 222). He *shouldn't* have accepted this answer—something does cause the star to stop.[41] But what Lewis wanted to say about this example applies to uncaused events generally. He was happy to say that uncaused events can be explained by the fact that they are uncaused. To keep things simple let's suppose that stellar collapse really is this kind of example. Here is how Lewis puts his response:

> I reply that information about the causal history of the stopping has been provided, but it was information of an unexpectedly negative sort. It was the information that the stopping had no causes at all, except for all the causes of the collapse which was [sic] a precondition of the stopping. Negative information is still information. If you request information about arctic penguins, the best information I can give you is that there aren't any.   (pp. 222–3)

Lewis's reply is a good one if the fact that the stopping was uncaused was offered (merely) as a *partial answer* to the question why the collapse stopped. But his reply is irrelevant, as a defense of (T0) anyway, if the fact that the stopping was uncaused was offered as a *reason why* the collapse stopped. The fact that the stopping was uncaused is not itself a cause of the stopping. So if this fact is a reason why the collapse stopped, we have a

---

[41] I defend this claim in "Are there Non-Causal Explanations (of Particular Events)?"

counterexample to (T0)—even though Lewis is right that the fact that the stopping was uncaused constitutes a partial answer to the question why the collapse stopped.[42]

But uncaused events are not, in fact, counterexamples to (T0)—for I see no reason to accept that the fact that E was uncaused is ever a reason why E happened. To whatever extent "E was uncaused" is a good thing to say in response to the question why E happened, it is only because it constitutes a partial answer to that question. "E was uncaused" is a way to convey that there *are no reasons* why E happened, not a way to convey that the fact that E was uncaused is itself a reason why E happened.[43]

Without the distinction between merely partial answers, and partial answers that also put forward reasons why E happened, Lewis is unable to distinguish the stronger reading of the example (where the fact that E was uncaused is offered as a reason why) from the weaker reading, or to disarm the stronger reading.

If we do not make use of the concept of a reason why we blind ourselves, we leave ourselves grasping in the dark for distinctions that are otherwise easy to see. Some philosophers invent new locutions for those distinctions, but the reasons-why terminology is better. In his paper "What Makes a Scientific Explanation Distinctively Mathematical?" Marc Lange argues that

> some distinctively mathematical explanations, though non-causal, nevertheless happen to cite the explanandum's causes [the causes of the event being explained]. Even so, they qualify as non-causal because they do not derive their explanatory power from

---

[42] Was the fact that the stopping was uncaused offered as a reason why? If you look back at Lewis's wording of the example, he uses "because": the star stopped collapsing "because it's gone as far as it can go . . . It's not that anything caused it to stop." We tend not to use "because" when we're giving partial answers to why-questions. Anyway, we can ask whether the example is a problem for (T0) on this reading, whether or not this is the intended reading.

[43] At the beginning of this section, when I was trying to figure out what kind of answer to a why-question Lewis's theory is a theory of, I moved very quickly past the hypothesis that it is a theory of reasons why. All I said was that he never used the term "reason why." I had another reason for rejecting that hypothesis, but we can only now appreciate it. Interpreting Lewis to have given a theory of reasons why yields a theory of reasons why that has little going for it and is not worth discussing in any detail. Here is the theory you get: the fact that R is a reason why E happened iff the proposition that R is about the causal history of E. If E is any uncaused event, then this theory entails that the fact that E was uncaused is a reason why E happened; which it is not.

their success in describing the world's network of causal relations specifically.   (p. 495)

I will discuss what Lange means by "distinctively mathematical explanation" in section 5.1. For now I want to focus on Lange's claim that some explanations that cite causes are not causal explanations. How can this be? If proposition P describes some causes of E, then according to Lewis's theory, and according to my theory, P is a partial answer to the question why E happened. In what sense, then, is P not a "causal explanation"? Does Lange reject our theories of partial answers? It is hard to say; Lange does not organize his discussion around the notion of an answer to a why-question. Instead Lange's focus is on "explanations." He says that an explanation that cites causes may not "derive its explanatory power" from the fact that it cites causes. Such an explanation "just happens" to cite causes, and "[the fact that] a distinctively mathematical explanation happens to cite facts about the explanandum's causes [causes of the event being explained] does not mean that it works by virtue of describing the explanandum's causes" (p. 496). Lange wants to say that citing causes only makes an explanation into a causal explanation if the fact that it cites causes is the "source" of that explanation's "explanatory power."

How is this talk of the source of an explanation's explanatory power to be understood? It is much easier to appreciate what Lange is trying to get at if his claims are expressed as claims about answers to why-questions, specifically as claims about reasons why. The first thing to do is to rewrite the sentence

> An explanation of E can happen to cite causes of E without working by virtue of citing those causes.

as

> An answer to the question why E happened can mention some of E's causes, without being a correct answer in virtue of the fact that it mentions some of E's causes.

Does this make sense? It always helps to think about questions other than why-questions. Is there any sense to be made of this claim?

> An answer to the question who came to the party can mention Saturn's rings, without being a correct answer in virtue of the fact that it mentions Saturn's rings.

It is not obvious what to think about this claim. But this much is certainly true: "Tim came to the party, and James Maxwell first figured out the nature of Saturn's rings" cannot be a correct answer to the question who came to the party in virtue of mentioning Saturn's rings—because it cannot be a correct answer to the question in the first place! Instead it is a body of fact consisting of two parts, an answer to the question, and also an irrelevant fact.[44]

One might think that it is always like this for why-questions about events. Any attempt to present a body of fact that (i) is an answer to the question why E happened, (ii) mentions a cause C of E, but (iii) is not correct in virtue of mentioning C, is really not an answer at all; it is at best the conjunction of an answer with the (irrelevant) fact that C happened.

But in fact this conclusion is wrong. For it could be like this: C is a cause of E; and, moreover, C is a reason why E happened—and so the occurrence of C may be mentioned in an answer to the question why E happened; but *the fact that C is a cause of E is not the reason why* ⟨*C is a reason why E happened*⟩. *This,* I think, is the distinction Lange was reaching for. Speaking in terms of reasons why, rather than "sources of explanatory power," provides a clearer way to draw it.

One of Lange's goals is to establish the existence of "non-causal explanations." He has his own definition of this term, which I will not discuss; what I care about is whether any of his examples are counterexamples to my theory, (T1). All I want to say about this now is that I do need to worry about his examples. If there is an example of the kind I am interpreting Lange to be describing, a truth of the form "that C is a reason why E happened" where the fact that C is, first, a cause of E, but second, is not a reason why E happened because it is a cause, then (T1) does stand refuted, for it says not just that all causes are reasons, but that all causes are reasons

---

[44] It's not true, however, that references to Saturn's rings are never relevant to answering the question who came to the party. I might answer this question with "That guy who is obsessed with Saturn's rings." Reference to Saturn's rings is here a means to identifying someone who came to the party. This returns us to the question of whether this answer is correct *in virtue of* mentioning Saturn's rings; I am not sure. (Maybe it is just correct in virtue of mentioning Ed, who just happens to be obsessed with Saturn's rings.) Anyway, the analogous answers to why-questions are not of interest. For Lange did not have in mind answers to the question why E happened that mention a cause of E merely as a means to refer to some other reason why E happened.

because they are causes. I will examine Lange's arguments that there are such examples in chapter 5.

I said that the biggest flaw in Lewis's theory is that it is incomplete. The biggest flaw lies not in what the theory says but in what it doesn't say: it says nothing about reasons why. This kind of flaw might not sound so bad in the abstract, until one remembers the damage that telling only part of the truth can do in a court of law. I have tried to illustrate the damage it does in this case. But still, isn't this flaw easily fixed? To make Lewis's theory complete we only need to revise his theories of partial and complete answers to speak in terms of reasons why rather than causes (so the theory of complete answers should say that the complete answer to the question why E happened lists all of the reasons why E happened), and add in (T0), the claim that the reasons why E happened are the causes of E. The theory we get is false—facts about what grounds E are also reasons why E happened—but at least it is not hiding anything and we know where to attack it.

If adding (T0) makes Lewis's theory complete, then, since I have also said that (T0) should to be understood to contain the same definitions of partial and complete answers that the amended version of Lewis's theory contains, maybe I should not have characterized Lewis's theory as an alternative to (T0), as I did at the beginning of this section. Yet still . . . there is some evidence that Lewis would not have accepted my contention that his theory is missing something.

At the end of his paper David Lewis compares his theory to the DN model. He sees something right about Hempel's theory: "if explanatory information is information about causal histories, as I say it is, then one way to provide it is by means of D-N arguments." But his interest in reconciliation only goes so far, and he goes on to identify what he takes to be a mistaken presupposition of the DN model. It is worth letting Lewis make his case at length, in his own words:

> The D-N argument . . . is represented as the ideal serving of explanatory information. It is the right shape and the right size. It is enough, anything less is not enough, and anything more is more than enough.
>
> Nobody thinks that real-life explainers commonly serve up full D-N arguments which they hope are correct. We very seldom do. And we seldom could—it's not just that we save our breath by leaving out the obvious parts. We don't know enough. Just try it. . . .

Hempel writes "To the extent that a statement of individual causation leaves the relevant antecedent conditions, and thus also the requisite explanatory laws, indefinite it is like a note saying that there is a treasure hidden somewhere." The note will help you find the treasure provided you go on working, but so long as you have only the note you have no treasure at all; and if you find the treasure you will find it all at once. I say it is not like that. A shipwreck has spread the treasure over the bottom of the sea and you will never find it all. Every doubloon you find is one more doubloon in your pocket, and also it is a clue to where the next doubloons may be. You may or may not want to look for them, depending on how many you have so far and on how much you want to be rich.

If you have anything less than a full D-N argument, there is more to be found out. Your explanatory information is only partial. Yes. *And so is any serving of explanatory information we will ever get*, even if it consists of ever so many perfect D-N arguments piled one upon the other. There is always more to know. A D-N argument presents only one small part—a cross section, so to speak—of the causal history.... A D-N argument might give us far from enough explanatory information, considering what sort of information we want and what we possess already. On the other hand, it might give us too much....

Is a (correct, etc.) D-N argument in *any* sense a complete serving of explanatory information? Yes in this sense, and this sense alone: it completes a jointly sufficient set of causes.... The completeness of the jointly sufficient set has nothing to do with the sort of enoughness that we pursue. There is nothing ideal about it, in general. Other shapes and sizes of partial servings may be very much better—and perhaps also better within our reach.

It is not that I have some different idea about what is the unit of explanation. We should not demand a unit, and that demand has distorted the subject badly. It's not that explanations are things we may or may not have one of; rather, explanation is something we may have more or less of.   (pp. 236–8; italics in original)

Lewis begins with an eloquent statement of the idea that we never traffic in complete answers to why-questions, that it is partial answers that we really care about. I remain skeptical. If "why" is a quantifier, if "Why did E happen?" is close in meaning to "Tell me all the reasons why E happened," then in typical contexts the domain of quantification will be restricted, and "tell me all the reasons why E happened" will mean something like, tell me all the reasons *of kind K* why E happened. So in a restricted context the complete answer to the question expressed by "Why did E happen?" might well be quite short, so short that we do "know enough," and have enough breath, to state it.

But this isn't the main point I want to make about what Lewis says.[45] The moral he draws at the end, about the unit of explanation: Lewis was wrong. Yes, it was a mistake to think that the DN argument was the basic unit of explanation. Yes, the assumption that "explanations are arguments" did distort the search for a theory of answers to why-questions.[46] But it is false to say that this mistake and distortion stemmed from a deeper mistake, that of thinking that there is a "basic unit of explanatory information." There is a basic unit. Hempel just had a false view about what it is. A complete answer to the question why Q is a conjunction of facts expressed by sentences of the form "that R is a reason why Q." The basic unit out of which answers to why-questions are ultimately built is the reason why.

The word "partial" contains "part," but there is a sense in which being a partial answer to a why-question is not sufficient for being *part* of the complete answer. The complete answer is a long conjunction, so one natural sense to give "part of the complete answer" has it that the parts of the complete answer are its sub-conjunctions: a (proper) part of the complete answer to the question why Q is a conjunction of some but not all of the facts expressed by sentences of the form "that R is a reason why Q." The atomic parts of complete answers are individual reasons why. "That Jones was at school is not a reason why the window broke" is a partial answer to the question why the window broke, but it is merely a partial answer, a partial answer that is not part of the answer. After learning it you know a little bit more about the complete answer, but still cannot identify a single one of the reasons why the window broke.

The metaphor of "explanation as treasure" is wonderful, and the way Lewis takes it from Hempel's hands and turns it against him is wonderful.

---

[45] I suspect he would have agreed with the point about domain restriction.

[46] Wesley Salmon called the thesis that all explanations are arguments a "third dogma of empiricism" in a paper with that title. (Many theses have been labeled *a* third dogma of empiricism—Davidson uses the label in "On the Very Idea of a Conceptual Scheme" for a different thesis than Salmon's—but none of them seems to have managed to become *the* third dogma of empirism.)

In philosophy metaphors can be powerful, but they can also be dangerous, and Lewis did not appreciate how dangerous this metaphor was for his own view. He did not have the metaphor completely under control. We should not demand a unit of explanation? But the treasure in Lewis's story does come in "units." A diver gathers the treasure *doubloon by doubloon*. Putting a doubloon in your pocket does not correspond to learning a (merely) partial answer to the why-question one has asked; for to acquire a doubloon is to acquire *part of* the treasure, not to "merely partially acquire" the treasure, whatever that might mean. Putting a doubloon in your pocket corresponds to learning a reason why. If, when searching for treasure, the best we can hope for is to have some but not all of the doubloons, if this is all we really care about anyway, and if answering why-questions is like treasure-hunting, then there is a "unit of explanation"—I claim that the reason why is the unit—and what we care about is not just learning some propositions that rule out one or another hypothesis about what the reasons why Q are, as Lewis maintained—for that *would* be like having only the note but no treasure at all—, but actually learning what some of those reasons are.

3.4 Preliminary Objections

Some "counterexamples" to (T1) are easier to handle than others. This section is for a few easy ones.[47]

A body exists alone in a Newtonian universe, moving along a straight line at constant speed. Why is it doing that, instead of accelerating? Because there are no forces acting on it. That is, the reason why it is not accelerating is that no forces are acting on it. But the absence of forces does not appear to be a cause; surely absences cannot be causes.[48] If this is right, then (T0) is false. Presumably the example, if good, refutes (T1) as well, since the absence of forces does not ground the body's inertial motion.

---

[47] I discussed several examples that have been offered as examples of "non-causal explanations" in "Are There Non-Causal Explanations (of Particular Events)?" My views have changed a lot since I wrote that paper, but I still stand by some of what I said. I am not going to rediscuss every example from that paper.

[48] Robert Brandon and, following him, Marc Lange, discuss this example ("The Principle of Drift," p. 321; "What Makes a Scientific Explanation Distinctively Mathematical?," p. 493).

This argument rests on controversial premises about causation. Two of them are: (i) fundamentally speaking, the causal relation relates events, and (ii) there are no such events as "absences." I reject (i). Since facts can be causes, the fact that there are no forces is available. But the argument would not convince me if someone convinced me that (i) is wrong. If at bottom causation is a relation between events, I am willing to embrace the existence of absences. There's a lot of metaphysics that needs to be done to defend one or another of these claims, but others have defended them well, I think; this is not the place.[49]

Other examples rest on different controversial theses about causation. Many philosophers hold that "dispositions explain their manifestations." For example: a wine glass is struck, and breaks; one reason why it broke is that it was fragile. But, the argument goes, dispositions do not cause their manifestations. That the glass was fragile was not a cause of its breaking. This is a controversial claim about causation, one that I do not find plausible.[50]

These two examples—one about inertial motion, the other about dispositions—were not intended by their authors to be counterexamples to (T0).[51] One author, Marc Lange, intended them, instead, to establish

[49] See for example Bennett, *Events and Their Names*; Mellor, "For Facts as Causes and Effects"; and Paul and Hall, *Causation: A User's Guide*.
  In "Causing and Nothingness" Helen Beebee argues that "we do not need absences as causes in order for facts about absences to be the explananta [things that 'do the explaining'] of causal explanations" (p. 301). About Brandon's example she would say that "the body moved at a constant speed because no forces acted on it" is true (and is a "causal explanation"), even though there are no such events as absences (of forces) to cause it to move at a constant speed. To back up her claim that "because no forces acted" is true she appeals to Lewis's causal theory of explanation, according to which this constitutes a partial answer to the question why the body moved at constant speed. Beebee would be right that we do not need absences as causes if facts about absences were only ever offered as partial answers to why-questions about events. But here, in this case, the fact that no forces acted is offered as a reason why the body did not accelerate—it is an instance of the "good strategy" I described in the previous section. This is not consistent with (T0) unless the absence of forces is a cause.

[50] A partial list of this argument's advocates: Lange, "What Makes a Scientific Explanation Distinctively Mathematical?," p. 494; Jackson and Pettit, "In Defense of Explanatory Ecumenism"; Hawthorne and Manley, Review of Mumford's *Dispositions*, p. 192. See section 6 of Fara and Choi, "Dispositions," for a survey of responses to the claim that dispositions are not causes.

[51] Of course the authors had never heard of (T0); I have formulated it for the first time here in this book.

that "causal explanations" need not cite causes. Was he at least right in his intended conclusion?

So far I have studiously avoided, as far as I could, using the terms "causal explanation" and "non-causal explanation." The time has come for me to say something about them. This game, of saying what it takes to be a "causal explanation," while popular, baffles me. I have discussed several kinds of answers to why-questions: partial answers, complete answers, and reasons-why answers. When a philosopher urges that causal explanations are explanations that meet conditions X, I never quite know what to think. The first place I get stuck is in figuring out which kind of answer the thesis is supposed to be about. Of course it is easy to try the various possibilities. But I run into more problems when I do. Let's try it for partial answers. Let's contemplate assertions of the form: a partial causal answer to the question why E happened is an answer that meets conditions X. How am I supposed to evaluate a thesis like this? How does a philosopher who proposes the thesis think I am supposed to evaluate it? He might think that the notion of a "partial causal answer" to a why-question is an ordinary notion, a notion that most non-philosophers make use of at least once in a while. If it is then he can safely assume that when his readers are first exposed to his arguments those readers already have opinions about what examples are, and are not, examples of partial causal answers. Those readers can then at least make a start on evaluating his thesis, by seeing whether the examples that are partial causal answers according to the thesis are the same as the ones the reader already believes are partial causal answers. (I am not saying that this is the only way to evaluate a philosophical thesis, just that, in many cases, it is a good place to start.)

The assumption that the notion one is theorizing about is an ordinary one is correct in many cases. An epistemologist is within her rights to make it when she proposes a theory of knowledge. She puts forward the view that S knows that P iff X; other epistemologists, equipped with pre-theoretic judgments about which cases are cases of knowledge, respond by checking whether the cases they have already judged to be knowledge are cases in which X. They do *not* respond by saying "wait a minute! What is this thing knowledge that you're proposing a theory of? Before you propose a theory of a phenomenon, you need to identify it, so your audience has some idea what you're talking about."

But that is how I would respond to a proposed theory of partial causal answers to why-questions. This is not an ordinary notion. It does need an introduction. If a team of graduate students were to go to Washington Square Park and say to passers-by "you're familiar with the notion of a partial causal answer to a why-question, right?", I doubt that anyone would have any idea what they are talking about.

An alternative to assuming that the notion of a partial causal explanation is an ordinary one is to say: "The term 'partial causal answer' is a technical term I am hereby introducing to do philosophical work." However, if that is what one is doing, then any theory of partial causal answers one went on to assert would be, to a large extent, not really a substantive theory but a stipulative definition.

This is the only sense I can make of theories of causal explanation. "Causal explanation" is not like "knows," an ordinary familiar notion. One cannot just dive headlong into providing a theory of it, assuming your readers know what you are talking about. All I get out of arguments that causal explanations are explanations that do this or that is that the purveyors of those arguments are stipulating a meaning for a technical term that has no established use outside of philosophy, and that, absent such a stipulation, is uselessly vague. From my point of view, their disagreements about what it takes to be a causal explanation are not substantive, they are just disagreements about how to use a bit of jargon.

I'm not saying that when I came across the term "causal explanation" for the first time (when I did, it was in a philosophy paper), I was completely unable to make judgments about which cases the term applied to. "The window broke because the rock hit it" should certainly count as a causal explanation. But the examples some philosophers use to reach finely-tuned conclusions about what is or is not a causal explanation—I don't know where their opinions about those examples are coming from.

Unlike most writers on the topic, Lange is refreshingly up-front about the status he thinks his theory of causal explanation has; he does not regard it as a theory of an ordinary familiar notion. When *he* argues that there are non-causal explanations, he writes,

> I am not appealing to some account of what makes an explanation 'causal' that aims to fit either some pretheoretic intuitions about which explanations are 'causal' or some scientific practice of labeling

certain explanations 'causal'. Rather, I am elaborating the notion of
'causal' explanation that is employed by those philosophers whom
I have quoted as contending that all scientific explanations are causal.

("What Makes a Scientific Explanation
Distinctively Mathematical?", p. 487)

There remain, however, two problems. First is that even if one is explicit
that one is stipulating a sense for "causal explanation," one has still not
specified whether this is a stipulated sense for "partial causal answer to
a why-question," "complete causal answer to a why-question," or what.
Second is that if one is introducing a technical notion, there is always
the worry that debates about whether that notion applies in a given case,
debates about whether an explanation is causal in the stipulated sense, are
not substantive debates. Why think that the stipulated notion of a causal
explanation is interesting, that the things that it applies to "go together,"
or form a "natural kind"? I think the place for substantive debate is over
whether (T0) or (T1) (or neither) is true; so I think that the debate over
whether some example is a causal explanation in some stipulated sense,
if that debate is not the same as the debate over whether the example is
consistent with (T0) or (T1), is not substantive.

I have said that when a philosopher jumps off the starting block and
says that causal explanations are those that do X, I can only regard this as
a stipulation. There is one exception, one way of proposing a distinction
between causal and non-causal answers to why-questions that doesn't
make me go blank. It is to propose a distinction between causal and
non-causal reasons-why answers to why-questions. If someone asked me
which reasons why E happened are the "causal" reasons why E happened,
I would still find the question a bit weird, but I would think there was
really only one answer: the causal reasons why E happened are the *causes*
of E.[52] So here is another place where I think philosophical debates over
the nature of explanation would benefit from focusing on reasons-why
answers to why-questions. With reasons-why on the table we can locate
the proper place for the debate over what it takes to be a "causal explanation," and hopefully even settle it.

---

[52] Technically my answer is more complicated: in line with (T1), it is that the causal
reasons why E happened are the facts that (i) are causes of the fact that E happened, and (ii)
are reasons why E happened because they are causes.

The "straight line at constant speed" example I started this section with is one in which a body is moving inertially, and the question is why it is moving inertially. Graham Nerlich added a twist to the example,[53] by considering a world in which space is curved, so that inertial motion is no longer motion in a straight line. (Inertial motion is still motion along a "straightest" line, but straightest lines in curved space are not straight.) He asks us to consider a cloud of dust moving inertially in non-Euclidean space. As it moves it changes its shape.[54] "Here," Nerlich writes, "is one example of an observably changing state of matter which involves no causes at all . . . No forces operate at all. The curvature of space explains the change of shape in the context of classical physics but, quite clearly, is nowhere causally involved in it" (p. 74). Up until the last sentence it looks like Nerlich is heading to the same place Brandon is: the claim that the cloud changed shape because no forces acted on it. But then he swerves. His actual claim is that the cloud changed shape because of the curvature of space. Making this explicitly a claim about reasons why gives

(6)   The fact that the cloud moved through a curved region of space is a reason why it changed shape.

But that it moved through a curved region did not cause it to change shape; we are working with Newtonian physics, in which space has no causal powers.

The first thing I want to say is that I do not think (6) is true. My reason is not that (6) would be a counterexample to my theory if it were true; even pre-theoretically it does not strike me as true. I think the fact that the cloud moved through a curved region of space is not specific enough to be a reason why the cloud changed shape. If we are to have a counterexample to (T1) we need to look around the neighborhood of (6) for something that is true. One candidate is

(7)   The fact that the cloud formerly occupied a region with one shape, and now occupies a region with a different shape, is a reason why it changed shape.

---

[53] So to speak; Nerlich's paper "What Can Geometry Explain?" was published much earlier than Brandon's.

[54] I think these are supposed to be special dust particles that do not exert forces on other dust particles.

This is not a counterexample to (T1), because the shapes of material things are grounded in the shapes of the regions of space they occupy.[55]

One might object: (7) is not exactly what Nerlich wrote; there are other why-questions about the cloud that Nerlich might have thought he was answering; and the reasons why those answers offer might not be grounds. Well, what are these other why-questions? We might ask why the cloud currently occupies *this* region of space. Answer:

> (8)   One reason why the cloud currently occupies *this* region is that it formerly occupied *that* one; another reason why is that no forces acted on it as it moved.

Now we have reasons why that are not grounds. But this time they are causes. (8) does not differ from Brandon's example in any interesting respect.

Maybe we still haven't latched on to the why-question Nerlich aimed at. After all, neither (7) nor (8) cites the curvature of space as a reason why something is the case, yet this seems to be what Nerlich had in mind. So here is a third try. Why is it the case that the region the cloud currently occupies differs in shape from the region it formerly occupied?

I think this question is most naturally read as having several presuppositions. Let's make them explicit. We are asking: *given that* the region the cloud currently occupies is the one you "get to" by moving a fixed distance along a collection of "straightest" line segments, all of which initially point in the same direction,[56] why does the current region differ in shape from the former region? One important reason why does seem to be the fact that the space through which those lines pass is curved. So here, finally, the fact that space is curved is serving as a reason why something is the case. And this fact is certainly not a cause of anything. Again, though, there is no threat to (T1). The sentence that follows "why is it the case that" in the question, namely "the region the cloud currently occupies differs in

---

[55] This is, I suppose, mildly controversial. But the alternative view, on which facts about the shapes of material bodies are just as fundamental as facts about the shapes of the regions they occupy, faces serious problems. See my "Are Shapes Intrinsic?"

[56] I put this claim in here because it seems to be a presupposition of the question. The question arises because the presuppositions create the expectation, for those who have Euclidean space in mind, that the two regions have the same shape, an expectation that is violated. But this particular presupposition is false, as Nerlich points out. There is no path-independent notion of "same direction" in curved space.

shape from the region it formerly occupied," does not "describe an event." It falls outside the scope of (T1) completely.

Bertrand Russell observed that "in advanced sciences such as gravitational astronomy, the word 'cause' never occurs," and went on to assert that "the reason why physics has ceased to look for causes is that, in fact, there are no such things" ("On the Notion of Cause," p. 1). How bad are things for my theory if Russell is right, and nothing has any causes? If Russell's thesis is right, then if my theory is also right, there are no reasons why anything happens (other than grounds—which will not matter here). This last claim certainly sounds wrong; certainly there are reasons why the ground is wet, the fact that it is raining among them. But maybe the claim that there are no reasons why anything happens only sounds wrong because we do not accept Russell's thesis. Maybe it would sound right, if we became convinced that Russell's thesis were true. "Look, you only think that the ground is wet because it's raining, because you think the rain caused the ground to be wet. But it didn't, since nothing causes anything." Maybe if people became convinced that Russell's thesis is true and heard a speech like this they would give up their belief that the ground is wet because it is raining. If so, things are not at all bad for my theory if Russell is right.

Luckily, the question of whether my theory is viable if Russell's thesis is true is not pressing, since very few people now accept Russell's thesis. (I will not take it upon myself here to argue that many events have causes.)

Another, weaker, thesis is suggested by what Russell says about the theory of gravitation:

> In the motions of mutually gravitating bodies, there is nothing that can be called a cause, and nothing that can be called an effect; there is merely a formula. Certain differential equations can be found, which hold at every instant for every particle of the system, and which, given the configuration and velocities at one instant . . . render the configuration at any later instant theoretically calculable . . . This statement holds throughout physics . . . But there is nothing that could be properly called "cause" and nothing that could be properly called "effect" in such a system.   (p. 14)

One way to put the point here is to say that there is no causation "in physics," or, better, in "fundamental physics." But this claim is obscure: either things have causes or they don't; if they do, then "there is causation"—full stop. What could it mean to say that there is causation, but not "in physics"?

One thing it could mean is that when you give a relatively complete description of some situation using the preferred vocabulary of a fundamental physical theory, then none of the events you've described cause any of the others. You write down a complete description of what Russell called "the configuration and velocities" of a bunch of particles at some time; you use the relevant differential equations to find the configuration and velocities at an earlier time; that earlier configuration, and those earlier velocities, do not cause the later configuration, and later velocities.

Some take this claim to be consistent with the claim that there is causation; they just say that the kinds of events that are causes, and are effects, are inexact, "high-level" events, events that can be realized in a variety of microphysically different ways. Throwing a rock at a window, we want to be able to say, is a way to break the window. And breaking the window is a way to cause the window to break. The idea is that the throw and the breaking are related as cause to effect only because there are many, many microphysically different events—differing over exactly where each of the atoms that make up the window is over some period of time—that can "realize" a window breaking, and similarly for a rock-throwing.[57] In this way, the kinds of events that cause and get caused resemble the properties that form the subject matter of thermodynamics. Temperature, pressure, volume, and so on, are magnitudes that can be realized in a variety of microphysically different ways. Hartry Field thinks this similarity in the circumstances in which causal talk, and in which thermodynamic talk, get their grip on reality, is not a coincidence: "The notion of causation, like the notions of temperature and entropy, derives its value from contexts where statistical regularities not necessitated by the underlying physical laws are important" ("Causation in a Physical World," p. 445).

This is a fascinating collection of ideas, and I have a lot of sympathy with some of them. Still, even though these ideas are weaker than Russell's thesis, they raise the same problem for my theory, since they contain the thesis that some events for which we think we have answers to the question of why they occurred lack causes, namely "exact, low-

---

[57] For this idea, see Field, "Causation in a Physical World"; Russell, in his paper, also emphasizes the importance of vaguely-specified events in our causal thinking.

level" events, like the event consisting in these particles being in this exact configuration. If no exact event has any causes, but there are still reasons why such events occur (that are not grounds), then (T1) stands refuted.

As Russell's thesis is, the claim that exact events lack causes is consistent with my theory, provided that the uncaused events also happen for no reason. Still, I do not think it is right to say that exact events lack causes. Suppose that at time T there are two massive bodies at rest ten meters apart, with nothing else around for miles (for light-years, even). Later, at time T+S, the bodies are closer together, say three meters apart, and moving towards each other at some specific velocities (which I won't bother specifying). Why are they closer together? The question has an answer, and the answer has something to do with the earlier positions of the bodies, and the gravitational forces between them. But the earlier positions of the bodies, and the gravitational forces the bodies exert on each other, also seem obviously to be causes of their present condition. So I reject the claim that there is no causation "in fundamental physics."

I do not think, though, that this rejection is very radical, or that it puts me very much at odds with those who think that Russell was on to something. The correct moral to draw, from thinking about fundamental physics and its use of differential equations, is not that causation is absent from fundamental physics but that our ordinary expectations about what the causes of an event are like are violated when we look at exact events as described in the language of fundamental physics. What are those ordinary expectations, and how are they violated? I imagine that an entire book could be written about these questions. Briefly, here are sketches of answers that I like. We ordinarily have expectations about how things of various kinds behave when nothing interferes with them—of what their default behavior is. When something exhibits its default behavior, there are no "positive" causes of its behavior, the only causes are facts about the absence of interference. When something deviates from its default behavior, on the other hand, we ordinarily think that there are a *small number* of *relatively localized* causes of this deviation.[58] Exact events as treated in fundamental physics violate these expectations. They do not

---

[58] These ideas are developed in many places, for example in Elga, "Isolation and Folk Physics"; see also the other papers in Price and Corry, *Causation, Physics, and the Constitution of Reality*. Maudlin argued that the default/deviant distinction is important to our causal thinking in "Causes, Counterfactuals, and the Third Factor"; many others have also taken up this idea.

have a small number of localized causes. The exact state of each part of the universe at some past time[59] is going to be a cause. If we focus on the small parts of the universe we will find localized causes, but there will be a very very large number of them. If we focus on the the universe as a whole we will get a small number of causes (in fact, one), but they will not be localized.

What is important about all of this here, though, is that an exact event *does* have causes, however different its causes may look from the causes of the kinds of events we are interested in in "the ordinary business of life" (or in sciences like meteorology).[60]

Bas Van Fraassen held that why-interrogatives are context-dependent: one and the same interrogative sentence can be used to ask different why-questions in different contexts of utterance.[61] I think he is absolutely right about this, and I think it is now a pretty common view. But Van Fraassen also traced the context-sensitivity of why-interrogatives to a particular source. He held that the question a why-interrogative asks in a context depends in part on what "relation of explanatory relevance" is salient in the context. Causation is, perhaps, one relation of explanatory relevance, but it is not the only one, and Van Fraassen seemed to think that there were lots and lots of relations of explanatory relevance, where "lots and lots" is certainly greater than two.

If Van Fraassen is right then (T1) is false, for (T1) is at best compatible with the thesis that causation and grounding are the only two relations of explanatory relevance. But the examples Van Fraassen used to motivate his view do not convince me. Here is part of his central discussion:

> Aristotle's lantern example (Posterior Analytics II, 11) shows that he recognized that in different contexts, verbally the same why-

---

[59] If we want to take the theory of relativity into account: the exact state of each part of the universe from which something traveling at or below the speed of light could reach the event in question.

[60] I do not intend these few paragraphs to be a comprehensive evaluation of the arguments that causation is "absent" from fundamental physics. I have not addressed, for example, the argument that our fundamental physical theories are time-symmetric, so there is no place for an asymmetric notion of causation in them. Mathias Frisch provides a comprehensive evaluation in his book *Causal Reasoning in Physics*, and argues that all the arguments fail.

[61] See *The Scientific Image*, chapter 5.

question [that is, the same interrogative sentence] may be a request for different types of explanatory factors. In modern dress the example would run as follows. Suppose a father asks his teenage son, 'Why is the porch light on?' and the son replies 'The porch switch is closed and the electricity is reaching the bulb through that switch.' At this point you are most likely to feel that the son is being impudent. This is because you are most likely to think that the sort of answer the father needed was something like: 'Because we are expecting company.' But it is easy to imagine a less likely question context: the father and son are re-wiring the house and the father, unexpectedly seeing the porch light on, fears that he has caused a short circuit that bypasses the porch light switch. In the second case, he is not interested in the human expectations or desires that led to the depressing of the switch. (p. 131)

Let us grant that "A reason why the porch light is on is that the switch is closed" is true in one context and false in the other (and "a reason why the light is on is that we are expecting company" is true in the other and false in the one). Van Fraassen's thesis is that this difference in truth-value is due to a difference in which relation of explanatory relevance is operative in the contexts; I interpret this to be the thesis that the word "reason" is context-sensitive, and has a different semantic value in the two contexts. I think his example supports a different thesis about the source of the context-sensitivity: contextual variation in the domain of quantification. "A reason why Q is that R" can be paraphrased roughly as "There is an X such that X is a reason why Q and X is identical to the fact that R." Our interests can influence what, in a given context, the quantifier "There is" has as its domain. Just as "There are five things in the fridge" can be true even though the refrigerator has millions of air molecules in it, because we have restricted our quantifier to edible things, "There is a reason why the porch light is on that is identical to the fact that the switch is closed" can be true in one context and false in another, even if "reason why the light is on" applies to the same set of causes of the light's being on in both contexts; all that is required is for the father through his interests to restrict the quantifier to certain causes, causes other than human intentions. Since domain restriction can account for the example, and since domain restriction is a well-attested phenomenon, there is no pressure, from this example anyway, to also say that "reason" is itself context-sensitive.[62]

Van Fraassen cares about the context-sensitivity of "the porch light is on because the switch is closed" because its context-sensitivity shows "Why is the porch light on?" to be context-sensitive. I hold the same view about the source of this interrogative's context-sensitivity. "Why is the porch light on?" is close in meaning to "Tell me all the reasons why the porch light is on,"[63] so "the porch light is on because the switch is closed" will be true (and so an answer) only in contexts where the causes/reasons in the domain quantified over by "all the reasons" includes the fact that the switch is closed.[64]

It may be that Van Fraassen marshaled a bad example to support a good thesis. For all I have said, "reason" in "reason why . . ." could still be context-sensitive. I am in fact prepared to be convinced that it is. Maybe in some contexts it applies only to causes, in others, only to grounds. What I deny is that it displays a tremendous amount of context-sensitivity. Since I know of no foolproof diagnostic test for context-sensitivity, and since (T1) is compatible with some context-sensitivity in "reason," I will leave this topic here.[65]

---

[62] I am here assuming an oversimplified model of contextual domain restriction. But what I said can be adapted to more realistic theories of domain restriction, for example that in Stanley and Szabó, "On Quantifier Domain Restriction."

[63] As I said earlier in connection with Lewis's claim that no one could ever give the complete answer to a why-question.

[64] One common way of thinking about the semantic values of interrogatives has it that the semantic value of an interrogative is a partition of the set of all possible worlds (a partition in turn can be taken to be a set of propositions that are exhaustive and pairwise incompatible). Each cell of the partition is a possible complete answer to the question, and the true complete answer is the cell in the partition that includes the actual world. (This proposal is originally due to Hamblin, in his 1958 paper "Questions." It is different from Karttunen's view, discussed earlier.) An interrogative is context-sensitive, on the Hamblin view, if its semantic values in different contexts are different partitions. This view of the semantic values-of-interrogatives is compatible with my claim about domain restriction. For the semantic-values-as-partitions view is silent about how, exactly, a single interrogative gets associated with different partitions in different contexts. One way this can happen is by variation in the domain of quantification over which "why," or "all reasons," ranges.

[65] (T1) would, of course, have to be reformulated to make it consistent with some moderate context-sensitivity in "reason." In Appendix D I will argue that "reason" in "reason why" is ambiguous between a meaning used to report reasons for action, and a meaning not so used; the question of whether "reason" is context-sensitive that I have been discussing is the question of whether it is context-sensitive when used with the second meaning, the only meaning I am using it with until we get to chapter 6.

4

Levels of Reasons Why

4.1 Equilibrium Explanations and a Puzzle

Theory (T1) fits with plenty of examples. But other examples fit less well, and one can hardly glance through a philosophy journal these days without stumbling across a paper proposing yet another example of a "non-causal explanation."

Let's start in 1983. In that year Elliott Sober published "Equilibrium Explanation," where he argued that "equilibrium explanations" are not "causal explanations." Sober's central example of an equilibrium explanation was R. A. Fisher's answer to the question why the ratio of males to females in humans is (currently) very close to 1:1.[1] "The main idea" of Fisher's answer, Sober reports, "is that if a population ever departs from equal numbers of males and females, there will be a reproductive advantage favoring parental pairs that overproduce the minority sex. A 1:1 ratio will be the resulting equilibrium point" (p. 201).

Really the main reason why the sex ratio among humans is very close to 1:1 is that humans are disposed to have sons and daughters in equal numbers.[2] That much is obvious; the harder question, the one Fisher really answered, is why humans have this disposition. Sober's summary of Fisher's answer is compressed; here is a longer version. Consider a

---

[1] Fisher, *The Genetical Theory of Natural Selection*. I follow Sober in simplifying what Fisher accomplished; Fisher actually answered the question why the amount of parental resources spent on male children equals that spent on female children. A sex ratio of 1:1 at the age of independence follows if male and female children are equally expensive. In fact the sex ratio in humans at birth is not 1:1, but is biased toward males, exactly because male children are less expensive (for more details see Sober, *The Nature of Selection*, pp. 51–7). For simplicity of exposition I will assume that in humans males and females are equally expensive.

[2] This is, of course, a chancy disposition, a propensity; as many parents know first-hand, equal numbers are not guaranteed, even with an even number of children.

hypothetical population in which almost everyone is disposed to produce more sons than daughters. In a population like that a couple's daughters will tend to have more children than their sons; for their daughters will have more potential mates available, and will face less competition for them, than their sons. So mutant couples who are disposed to have more daughters than sons will (tend to) have more grandchildren.[3] The disposition to have more daughters than sons is therefore the fitter trait in that environment. As natural selection operates the genes that code for that trait will become more common. It's the other way around in a hypothetical population in which the disposition to have more daughters than sons is prevalent: the disposition to have more sons than daughters is fitter. But when the disposition to have sons and daughters in equal numbers is prevalent, there is no fitness advantage to being disposed to overproduce one of the sexes.

We can put this in terms of how a *population's* sex-ratio disposition changes over time. Let us say that a population is disposed to produce sons and daughters in the ratio $n{:}m$ iff $n{:}m$ is the "average," in the most natural sense, of the dispositions to produce sons and daughters had by the members of that population. Then the disposition (of a population) to produce sons and daughters in equal numbers is a "stable equilibrium point," in the sense that (i) if a population gets very close to having this disposition, it stays very close (where a disposition to have sons and daughters in the ratio $n{:}m$ is close to the disposition to have them in equal numbers to the extent that n/m is close to 1), and (ii) if a population is far from having this disposition, but not too far, then its disposition moves toward the disposition to have sons and daughters in equal numbers.[4]

---

[3] In more mathematical terms: if N is the number of people in the "third generation," f the number of females in the second generation, and m the number of males in the second generation, then on average a female in the second generation has N/f children, and a male, N/m. If most couples in the first generation are disposed to have more sons than daughters, then $m > f$, so $N/f > N/m$—a daughter born to a couple in the first generation has, on average, more children than a son. (This form of the argument is originally due to Carl Düsing. See Sober, *Did Darwin Write the Origin Backwards?*, section 3.5, for more on Düsing's argument; Sober, in section 3.6, points out a subtle difference between Düsing's and Fisher's conclusions.)

[4] In this case the population's disposition does not have to be very close to be not too far away. In fact, in the mathematical model, as long as the population is not disposed to produce only daughters or only sons, its disposition will tend toward the equal disposition (though the mathematical model is idealized, and in reality chance events can derail this process).

Here is Sober's argument that this is not a "causal explanation":

> a causal explanation ... would presumably describe some earlier state of the population and the evolutionary forces that moved the population to its present configuration ... Where causal explanation shows how the event to be explained was in fact produced, equilibrium explanation shows how the event would have occurred regardless of which of a variety of causal scenarios actually transpired.
>
> (p. 202)

Later he expands on this claim: "Equilibrium explanations present disjunctions of possible causal scenarios; the actual cause is given by one of the disjuncts, but the explanation doesn't say which" (p. 204).

So, are equilibrium explanations consistent with (T1)? Let "Fisher's Information" be the body of fact that constitutes Fisher's response to the question why humans are disposed to have sons and daughters in equal numbers. Sober held that Fisher's Information does not cite any causes of humans' having this disposition. But my theory does not require a *partial answer* to the question why humans have this disposition to cite any particular causes of it. So if Fisher's Information is merely a partial answer to the question why humans are disposed to have sons and daughters in equal numbers, it is no threat to my theory.

Fisher's Information rules out plenty of hypotheses about the causes of humans being disposed to have sons and daughters in equal numbers. It rules out, for example, the hypothesis that the disposition to have more daughters and sons is the fitter trait in every environment (not just one in which there are more males than females in the population), but a series of flukes has conspired to foil the efforts of natural selection to spread this disposition through the population. If things had been like that, then the fact that humans are disposed to have sons and daughters in equal numbers would have been caused not (or not just) by natural selection but (also) by other forces.

Fisher's Information is more of a threat to my theory if it is more than a partial answer, if it does more than just rule out some hypothesis about the reasons why humans are disposed to have sons and daughters in equal numbers. It is a threat if it identifies some individual reasons why humans are disposed to have sons and daughters in equal numbers. For if it puts forward a fact F as a reason why humans are disposed to have sons and daughters in equal numbers, and if Sober is right that Fisher's Information does not describe any causes, then F is a reason why that is not a cause.

There are two ifs here. *Is* Fisher's Information more than a partial answer? We could try to directly evaluate whether any of the propositions that constitute that information are reasons why humans are disposed to have sons and daughters in equal numbers. To do this we could choose a proposition p from those propositions, attend to the claim that p is a reason why humans are disposed to have sons and daughters in equal numbers, and try to decide whether this claim is true. I am later going to argue that most claims of this form are false—most of Fisher's Information consists of things other than reasons why humans are disposed to have sons and daughters in equal numbers. But first I want to focus on the second "if." Is Sober right that Fisher's Information does not provide any causes of humans' having this disposition?

Sober's argument for this thesis is bad. He assumes that the only relevant causes of the fact that humans are disposed to have sons and daughters in equal numbers are facts of the form *at time T humans were disposed to have sons and daughters in the ratio m:n*. Suppose that in the year 1000 humans had the "2:3" disposition. If Sober is right, this fact is the only relevant cause of humans currently being disposed to have sons and daughters in equal numbers that concerns the year 1000. I'm with those who reject this claim. Causes must be proportional to their effects. The intuitive idea of proportionality is that, other things being equal, a fact C is a cause of a fact E only if "nothing more than C was needed" to ensure that E obtained, and also "C itself was not more than was needed" to ensure that E obtained.[5] One way to spell this out is in counterfactual terms: C is a cause of E only if (i) there is no weaker fact C— with the property that, had C— obtained and C not obtained, E still would have obtained (C is not more than was needed); and (ii) every stronger fact C+ has the

---

[5] Preempting causes can fail to be proportional to their effects. Other things are not equal when there are preempted backups.

Stephen Yablo is a champion of proportionality; see "Wide Causation," p. 294, or "Mental Causation," section 8. Michael Strevens uses proportionality to argue that equilibrium explanations are causal explanations according to his theory in *Depth*.

A "pure" proportionality theorist says that proportionality is necessary for causation as long as preemption can be ignored. Neither Yablo nor Strevens is a pure proportionality theorist, nor is the pure version of the view plausible. Roughly speaking, both hold that when the potential cause that is proportional to the effect is "too disjunctive," it may lose the race to be the cause to a less proportional, but also less disjunctive, candidate.

property that, had C obtained and C+ not obtained, E still would have obtained (nothing more than C was needed).[6]

The fact that in 1000 humans had the 2:3 disposition is too strong to be a cause of the current sex ratio. It is "more than was needed." The weaker fact that in 1000 humans were not disposed to have only sons or only daughters is better proportioned to the effect. For had it been that in 1000 humans did not have the 2:3 disposition but still were disposed to have both sons and daughters, it still would have been that humans now were disposed to have sons and daughters in equal numbers.[7]

We are considering the following argument. Premise 1: Fisher's Information provides at least one reason why humans are disposed to have sons and daughters in equal numbers; premise 2: Fisher's Information provides no causes of this fact; so not all reasons why are causes. I have rejected Sober's argument for the second premise. But the premise could still be true, even if Sober had a bad argument for it.

Well, *does* Fisher's Information provide any causes? Sober, in his summary of Fisher's Information, never mentions facts like the fact—call it F—that in 1000 humans were not disposed to have only sons or only daughters, the facts that I have said are the relevant causes. Fisher does not mention them either, in his own presentation of Fisher's Information. This is pretty good evidence that facts like F are not part of Fisher's

---

[6] Proportionality need not be defined in terms of counterfactuals; Strevens, in *Depth*, does not do it that way.

I leave the notion of C+ being stronger than C deliberately vague. One could take C+ to be stronger than C if C+ asymmetrically entails C. Yablo prefers to use the notion of a "way for C to obtain," understood so that not every fact that asymmetrically entails C is a way for C to obtain.

Conditions (i) and (ii) are as Yablo states them in "Wide Causation" (p. 294). There are problems with clause (ii), which need not detain us; for the record, a better version of it reads like this: for every stronger fact C+, had C obtained because C+ did, E would still have obtained (Brad Weslake, in "Proportionality, Contrast and Explanation," suggests a similar improvement).

[7] The claim that an answer to the question why E happened that abstracts away from the nitty-gritty details of the causal process that produced E is a "non-causal explanation" is repeated by Batterman in, for example, "Multiple Realizability and Universality" (p. 28) and "On the Explanatory Role of Mathematics in Empirical Science" (p. 2). Batterman's picture is that abstracting from the nitty-gritty details takes you away from the causes of the event in question. But the proportionality requirement shows his picture to be false. In many cases abstracting away from the details take you *toward* the causes of the event; abstraction can take you away from overly-specific, and so out-of-proportioned, facts to less specific, and so better-proportioned, facts.

Information. On the other hand there is the fact that Fisher's Information is not a very good answer to the question why humans are disposed to have sons and daughters in equal numbers if it omits facts like F. Surely these facts are crucial! I think we should understand facts like F to be part of Fisher's Information, but facts that go without saying, since everyone already knows that they obtain.

Fisher's Information, then, does cite a cause of the fact that humans are disposed to have sons and daughters in equal numbers. Premise 2 is false. It has taken me a while to get to this conclusion.[8] Has all the work been for nothing? Even if Fisher's Information includes facts like F, it also includes so much more. And it seems absurd to say that those other propositions, while part of Fisher's Information, are not part of Fisher's answer, are not offered as reasons why humans are disposed to have sons and daughters in equal numbers. After all, everyone already knew that facts like F obtained (everyone already knew that humans have always been disposed to have both sons and daughters)—but they did not know why humans are disposed to have sons and daughters in equal numbers.

Let's attend more carefully to the argument this last sentence gestures at. The argument aims to refute the suggestion that

(1) The only parts of Fisher's Information that are offered as reasons why humans are disposed to have sons and daughters in equal

---

[8] Could I have gotten there sooner? Sober mentions that Fisher's argument requires the assumption that humans mate more or less at random, rather than preferentially choosing mates from "within one's group." When there is "inbreeding," the operation of group selection can produce a female-biased sex ratio (Sober presents the details in *Did Darwin Write the Origin Backwards?*, pp. 114–16). So isn't the fact that humans in the past were disposed to mate more or less at random a cause of humans' now being disposed to have sons and daughters in equal numbers? In the *Did Darwin* book (published in 2011) Sober seems to think it is; he calls "breeding structure" a "positive causal factor" (p. 116). In the equilibrium explanations paper, on the other hand, he seems to think it isn't. I'm not sure I follow the justification he offers (on p. 204); one possible interpretation is that he regards the random mating as a *ground* of the fact that being disposed to have sons and daughters in equal numbers is the fittest trait, and concludes from this that it cannot be a cause of that trait's becoming more prevalent. I think Sober was right in 2011, not in 1983. I suppose that the Sober of 2011 could say that the existence of this cause, the fact of random mating, does not automatically show premise 2 to be false, because it is not part of Fisher's Information, but is instead just a "presupposition" of that information. Or he could claim, as I am about to say in the main text, that Fisher's Information contains many reasons why humans are disposed to have sons and daughters in equal numbers other than this cause; so the argument can be fixed by using this claim as a premise. (I will say more about presuppositions of answers to why-questions in chapter 5, section 5.3.)

numbers are facts of the form *in year N humans were not disposed to have only sons or only daughters.*

Let Jones be a typical person who wanted to know why humans are disposed to have sons and daughters in equal numbers, and was provided with Fisher's Information. The argument that (1) is false goes like this:

(2)   Before learning Fisher's Information, Jones knew every (relevant) fact of the form *in year N humans were not disposed to have only sons or only daughters.*

(3)   Before learning Fisher's Information, Jones did not know why humans are disposed to have sons and daughters in equal numbers.

(4)   After learning Fisher's Information, Jones did know why humans are disposed to have sons and daughters in equal numbers.

(5)   Therefore, some part of Fisher's Information that is not a proposition of the form *in year N humans were not disposed to have only sons or only daughters* is a reason why humans are disposed to have sons and daughters in equal numbers.

This argument is not formally valid, but it still looks pretty good. To know why Q, one must know all of the relevant reasons why Q. So how could learning Fisher's Information result in Jones' knowing why humans are disposed to have sons and daughters in equal numbers, if that information did not provide Jones with the relevant reasons why humans are disposed to have sons and daughters in equal numbers? Assuming that it did provide him with those reasons, those reasons must be facts that Jones did not already know, so none of them can be a fact of the form *in year N humans were not disposed to have only sons or only daughters.* Conclusion: some other facts in Fisher's Information are reasons why humans are disposed to have sons and daughters in equal numbers. Since those other facts are not causes, they are reasons why that are not causes.

Seductive as this argument is, it is a bad argument. It can happen that you come to know, for the first time, why Q, after learning some facts F, without any of the facts in F being reasons why Q. This *can* happen; it is, I think, what *does* happen in the case of Jones.

How can it happen? Observe that I can know, say, that Smith is eating chocolate, and that he is nervous about his job interview, without knowing that he is eating chocolate *because* he is nervous about his job interview.

Just suppose that I do not know that eating chocolate calms his anxiety. So I can know a fact R that is a reason why Q, without knowing that R is a reason why Q. In a situation like this, if I find myself wanting to know why Q, I do not need to learn that R is true; what I need to do is "put two and two together," and learn that R is a reason why Q. Now if *you* know all this, and you have taken it upon yourself to help me learn why Smith is eating chocolate (maybe I asked you why he is eating chocolate), then (perhaps) the most economical thing for you to do is *not* to tell me that he is eating chocolate because he is nervous about his job interview, but instead just to tell me that eating chocolate calms his anxiety. By doing this you put me in a position to figure out on my own that he is eating chocolate because he is nervous.

I asked three paragraphs back "how could learning Fisher's Information result in Jones' knowing why humans are disposed to have sons and daughters in equal numbers, if that information did not provide Jones with the relevant reasons why humans are disposed to have sons and daughters in equal numbers?" The same way. Before learning Fisher's Information Jones knows, of the facts that are the reasons why humans are disposed to have sons and daughters in equal numbers, *that they are facts* (that they obtain). He does not know *that these facts are reasons*. What Fisher's Information does is put him in a position to figure this out.

We need to distinguish between, on the one hand, an answer to a question, and on the other, a *good response* to a question (in a given context). Obviously, if someone asks a question, one way to give a good response is to answer the question. But that is not the only way. Another way to give a good response is to provide the asker with information that puts them in a position to figure out the answer on their own.[9]

That there is a distinction between answers and good responses cannot be denied. If A asks "Is Connor coming to the party?" and B replies "He is sick," the sentence B utters does not express an answer to A's question. The only two possible answers are: he is coming (yes); he is not coming

---

[9] These are not the only ways to give a good response. "I don't know" and "You should ask George" can obviously, in some sense, be good responses to a question. But they are good responses only when the responder either does not know the answer, or has some reason not to provide the answer (say, he's promised George not to spread the news without permission). Here and throughout I am interested in non-answers that are good responses to questions even when the responder knows the answer and has no reason to hide it. I will discuss another example of a good response of this kind later in this chapter.

(no). B's reply is, however, certainly a good response. After hearing what B says, A knows the answer. In this case the response is good because it conversationally implies the answer. One of the maxims Grice lists in "Logic and Conversation" is the maxim of relation, which enjoins good conversational participants to say only what is relevant. "He is sick" is relevant only if B knows (i) that Connor is not coming, and (ii) that he is not coming because he is sick. Since A can assume B is obeying Grice's maxims (B has given no sign that he refuses to cooperate), A can work out that B knows these things, and so learn that Connor is not coming.

The distinction between an answer and a good response applies to why-questions as much as to any other kind. From the fact that providing a body of fact F is a good response (in context) to the question why Q, it does not follow that "Q because F obtains" is true. It does not follow that "Q because F obtains" expresses an answer to the question why Q.

Fisher's Information (or the part that goes beyond the fact that humans were never disposed to have only sons or only daughters in the past), I think, constitutes a good response to the question why humans are disposed to have sons and daughters in equal numbers, without constituting an answer to that question.

We can represent (the relevant bits of) Fisher's Information as a collection of subjunctive conditionals, conditionals of the form *if in year M humans had had the m:n disposition, they would now (still) have been disposed to have sons and daughters in equal numbers.* Putting these in your response to the question why humans are disposed to have sons and daughters in equal numbers can make that response good, even though they are not reasons why humans are so disposed, because learning that they are true can help one's audience come to know that facts of the form *in year M humans were not disposed to have only sons or only daughters*—facts they already know obtain—are (the only) reasons why humans are disposed to have sons and daughters in equal numbers. They can help one's audience come to know this, of course, because they can help one's audience come to know that these facts are better proportioned to the effect than more specific facts about past dispositions. To have a name for the role they are playing, we can say that the conditionals in Fisher's Information are "epistemic enablers."

In my argument that we need to distinguish giving a good response to a question from answering that question I looked at the

Connor-and-the-party example; there the good response that was not an answer was good because it conversationally implied the answer. The most conservative conclusion to draw from the example is that a response is good *only if* it conversationally implies an answer. It would be bad for me if this conclusion were right, for asserting the conditionals in Fisher's Information is not (in any ordinary context) a way to conversationally imply the answer to the question why humans are disposed to have sons and daughters in equal numbers. Fortunately, the conservative conclusion is too conservative. Saying something that conversationally implies the answer is *one* way to say something that puts one's audience in a position to know what the answer is. But it is not the only way, and putting one's audience in a position to know the answer seems like a good way to respond to a question regardless of the means one uses.

The distinction between epistemic enablers and reasons why is enough to show that Fisher's explanation, and equilibrium explanations in general, are compatible with (T0) and (T1). But I think that the relevant facts in Fisher's Information are not merely epistemic enablers; they are more. To isolate this additional role they play I need to bring on stage and formally introduce[10] the distinction that will play a starring role in the rest of this book.

4.2 Higher and Lower Levels of Reasons Why

For a given event E there are the reasons why it occurred. There are also the reasons why those reasons are reasons, that is, the reasons why those reasons are reasons why E occurred.

The iteration can make one's head spin. Let me go through it slowly.

An answer to the question why E happened says something about reasons of the first kind, reasons why E happened. But it is easy, whenever you entertain a proposition, to wonder why that proposition is true. So if you ask why E happened and are told that one reason why E happened is F, it is easy to wonder: why is it the case that F is a reason why E happened? An answer to *this* question says something about higher-level reasons, reasons why ⟨F is a reason why E happened⟩.

---

[10] It has already made a few informal appearances.

Requests for higher-level reasons occur naturally when the topic of conversation is a why-question. I was with my family at a children's museum not long ago. We had been to the museum several days in a row, and on our final visit my wife and I had the following conversation:

D: Why is the museum so much more crowded today than last time?

B: Because today is Friday.

D: What does that have to do with it?

B: I don't know, but that must be the reason.

My wife's follow-up question, "What does that have to do with it?", was a way of asking why the fact that today is Friday is a reason why the museum is so crowded. She was asking for a higher-level reason. (I, unfortunately, did not know.)

Here is another example of a follow-up why question and a higher-level reason; it is a bit artificial, but I will use it to make a point. I will call it

FIRST ROCK DIALOGUE:

A: Why did that rock hit the ground at a speed of 4.4 m/s?

B: The reason why it hit the ground at that speed is that it was dropped from a height of one meter.

A: Whoa! I don't understand. Why is it that its being dropped from that particular height is a reason why it hit the ground with that particular speed?

B: Because it is a law that the speed of impact, s, is related to the distance fallen d by the equation $s = \sqrt{2dg}$ (where g is the gravitational acceleration near the surface of the earth); and $\sqrt{2 \cdot 1 \cdot 9.8} \approx 4.4$.

In FIRST ROCK DIALOGUE the fact that the rock was dropped from one meter is offered as a reason why it hit the ground at 4.4 m/s, while the law that $s = \sqrt{2dg}$ is offered as a second-level reason why, a reason why the drop height is a reason why the impact speed is 4.4 m/s. The law shows up in the answer to the second-level why-question, not in the answer to the first-level one.

The mention of laws in answers to why-questions brings Hempel's DN model of explanation to mind, since that model requires all answers to why-questions to include laws. Keep the DN model in the background for just a minute, though. There is a general thesis I want to advance first, a thesis about the relationship between higher- and lower-level reasons why:

HIGH↛LOW: The following conditional is false: necessarily, for any A, B, and C, if A is a reason why ⟨B is a reason why C⟩, then A is a reason why C. That is, the proposition that A is a reason why ⟨B is a reason why C⟩ does not entail the proposition that A is a reason why C. Second-level reasons are not automatically first-level reasons.

Thesis HIGH↛LOW is an important part of my theory. It will play a large role in what I have to say about many alleged examples of "non-causal explanations."

It is easy to prove that HIGH↛LOW is true. Suzy throws a rock at a window but Billy sticks his mitt out, thereby catching the rock before it hits. The fact that Billy stuck his mitt out is a reason why the window didn't break. And the fact that Suzy threw the rock is a reason why ⟨the fact that Billy stuck his mitt out is a reason why the window didn't break⟩. But the fact that Suzy threw the rock is *not* a reason why the window didn't break.

All it takes to establish HIGH↛LOW is one example like this, for HIGH↛LOW is quite weak. It should not be confused with the much stronger, and definitely false, thesis

LEVEL SEPARATION: Necessarily, if A is a reason why B is a reason why C, then A is not a reason why C.

Thesis HIGH↛LOW says that if "A is a reason why B is a reason why C" is true, "A is a reason why C" *can* be false; LEVEL SEPARATION, in contrast, says that if "A is a reason why B is a reason why C" is true, "A is a reason why C" *must* be false. Examples of "joint causation" falsify LEVEL SEPARATION. I strike a match and it lights. One reason why it lit is that I struck it. Why is this a reason? Because oxygen was present. That is: the fact that oxygen was present is a reason why ⟨the fact that I struck the match is a reason why it lit⟩. But the presence of oxygen is *also* a first-level reason. It is also itself a reason why the match lit.

In fact in this case, each fact—the fact that I struck the match, and the fact that there was oxygen in the room—is a first-level reason why the match lit, and also a second-level reason why the other fact is a first-level reason why the match lit. Each is a reason why the match lit because each is a cause. Each is a reason why the other is a reason why the match lit because each requires the other to obtain in order to make the match light.

As an aside, these claims are a chance to say something about the distinction between the causes of an event and the "background conditions" to its occurrence. A paradigm case of a background condition is the presence of oxygen in the room when the match is struck. Intuitively, background conditions are conditions that enable causes to produce their effects. It is common for philosophers to think that the distinction between causes and background conditions is not very deep. It is common for them to hold that background conditions are just more causes. This is the view I was assuming when I said that the presence of oxygen was, not just a reason why ⟨the striking of the match is a reason why it lit⟩, but is also itself a cause of the lighting. When discussing some given event, we ignore some of its causes, or take their occurrence for granted, thereby making it the case that the word "cause" does not apply to them in that conversational context. But, speaking unrestrictedly, the events that are "merely background conditions" in that context are still causes.

Still, one cannot help but sometimes wonder, I myself sometimes wonder, if there is something deeper to the distinction between causes and background conditions than this. The distinction between levels of reasons why provides one way to make sense of the view that, even when the context places no restrictions on the extension of the word "cause," background conditions are not causes. The view could be that a background condition is a event, the occurrence of which is a reason why some other event C is a cause of E (and so also a reason why the occurrence of C is a reason why E happened), without also being itself a cause of E.[11]

---

[11] Dretske comes close to this theory of background conditions (*Explaining Behavior*, pp. 39–41). There are two differences: first, although he flirts with the idea, he officially declines to require background conditions to the occurence of E to *not* be causes of C; and second, he does not use the reason-why terminology, instead saying that a background condition "is a cause of C's causing E" (p. 40). I'm not sure whether the presence of oxygen really does count as a background condition on this view, because I'm not sure whether the presence of oxygen was a cause of the fact that ⟨the striking of the match caused the lighting of the match⟩.

When X is "a cause of C's causing E," Dretske calls X a "structuring cause of the process C → E" (pp. 42–4). He contrasts structuring causes with triggering causes; a triggering cause of the process C → E is just a cause of C. I'm not sure I believe in the existence of structuring causes, so defined.

I am more inclined to believe in them if "structuring cause of the process C → E" is defined as "cause of E that is also a reason why C is a cause of E"—it's worth noting that Dretske says that questions of the form "Why did C cause E?" ask for structuring causes—but I'd have to think harder about what it is to be a process. If, whenever C is a cause of E, there is a process C → E, I'd have to say that the striking of the match is a structuring cause of the

A big challenge for this theory of background conditions, of course, is to get it to give the right results about particular cases. Could it really be true that the presence of oxygen is a reason why ⟨the striking of the match caused the match to light⟩, but is not itself a cause of the lighting? The dominant theories of causation have it that the presence of oxygen *is* a cause of the lighting.[12]

As a second aside, the match-lighting example is relevant to the evaluation of the unrestricted version of (T1)—theory (T2)—that I mentioned in the last chapter. That theory says that, no matter what fact you are asking about, whether it corresponds to the occurrence of a concrete event or not, the reasons why that fact obtains are all and only its causes and grounds. The match-lighting example is, I think, a counterexample to (T2). The fact that oxygen was present is a reason why ⟨the fact that the match was struck is a reason why it lit⟩; but the fact that oxygen was present is certainly not a cause of the fact that ⟨the fact that the match was struck is a reason why it lit⟩ (it is not the sort of fact that can have causes[13]), and I do not think it grounds the fact that the striking is a reason why the match lit either.

So, to get back to the main thread, thesis HIGH ↛ LOW should not be confused with the stronger thesis LEVEL SEPARATION. Neither should it be confused with the thesis that "reason why" is not transitive:

NON-TRANSITIVITY: The following conditional is false: necessarily, for any A, B, and C, if A is a reason why B, and B is a reason why C, then A is a reason why C.

I think that NON-TRANSITIVITY, like HIGH ↛ LOW, is true. I think that causation is not transitive; since I think that causes are reasons why, any

---

process *oxygen present → match lights*, which of course I don't want to say. But, intuitively, there does not seem to be any such process as this. (I should say that Dretske seems not to always use the "structuring/triggering" terminology in the same way. In "Triggering and Structuring Causes" he distinguishes between triggering and structuring causes, not of processes, but of events. Moreover, he says at one place that the structuring causes of E are not the "standing conditions in which C causes E"—which I would call reasons why C caused E—but the "triggering causes" of the presence of those standing conditions (p. 140).)

[12] For example, most counterfactual theories of causation take counterfactual dependence to be sufficient for causation, and the counterfactual "If there had been no oxygen, the match would not have lit" is true.

[13] That is: the fact that ⟨the fact that the match was struck is a reason why it lit⟩ does not correspond to the occurrence of an event. I trust this is not a controversial claim.

example of causation's failure to be transitive shows NON-TRANSITIVITY to be true.[14]

One reason why it is easy to confuse HIGH↛LOW and NON-TRANSITIVITY is that the same example shows both to be true, the Suzy-Billy-window example. The fact that Suzy threw is not just a reason why (the fact that Billy stuck his mitt out is a reason why the window didn't break); the fact that Suzy threw is also a reason why Billy stuck his mitt out (since the throw caused him to do this—he wasn't sticking his mitt out in random disregard of his surroundings). But again, the fact that Suzy threw is not a reason why the window broke. Since Billy's sticking out his mitt *is* a reason why the window broke, NON-TRANSITIVITY is true. Despite this, once again, HIGH↛LOW and NON-TRANSITIVITY are distinct theses.

HIGH↛LOW and NON-TRANSITIVITY are both theses about answers to follow-up why-questions. As anyone with children knows, whenever you answer a why-question, you create an opportunity for your questioner to immediately ask "why?" about your answer. But there are two importantly different ways to ask "why?" about the answer to a why-question. On the one hand, you can ask *why the fact offered as a reason obtains*. Someone says "Q because R" and you ask "Why is it the case that R?" This is a "horizontal" (as we may call them) follow-up why-question. For any fact F you start with, there are the reasons why F obtains, and then there are the reasons why those reasons themselves obtain, and so on. By asking horizontal follow-up why-questions one traces this chain of reasons "backward" from F. This may be easier to understand if we assume, just for illustration, that all reasons are causes. Then for any event you start with, there are its causes, and the causes of its causes, and so on, and by asking horizontal follow-ups one traces this chain of causes back toward the beginning of time. Why is the window broken? Because Suzy threw a rock at it. Why did she do that? Because Billy has been ignoring her again. Why is he ignoring her? And so on, and on. NON-TRANSITIVITY is a thesis about answers to horizontal follow-ups.

But horizontal follow-ups are not the only kind of follow-up why-question. One may also ask *why the fact offered as a reason is a reason*. This is a "vertical" follow-up. By asking it, we step outside the chain of reasons that NON-TRANSITIVITY concerns. Instead of asking what facts

---

[14] One standard argument that causation is not transitive may be found in Hitchcock, "The Intransitivity of Causation Revealed in Equations and Graphs."

belong in the chain, we are now asking what the facts in the chain have done to belong in the chain. We are asking, not which facts are reasons, but why those facts are reasons. Why is the window broken? Because Suzy threw a rock at it. Why did she do that? Because Billy has been ignoring her again. *What does that have to do with it?* There is the vertical follow-up: instead of asking why Billy has been ignoring Suzy, we have in effect asked why the fact that Billy is ignoring her is a reason why she threw a rock. HIGH↛LOW is a thesis about vertical follow-ups.

The concept of a reason why one thing is a reason why another thing happened is closely related to the concept of a reason why one thing is a *cause* of another thing (this relationship will play a big role in the next chapter). And this second concept is closely related to Stephen Yablo's concept of an "ennobler" (as introduced in "Advertisement for a sketch of an outline of a proto-theory of causation," pp. 98–9). Ennoblers, he says, are facts without which the cause would not be required for the effect. He gives an example with the same structure as the Suzy-Billy-window example; about my example he would say that Suzy's throw ennobles Billy's sticking out of his mitt, since without her throw, the window's survival does not require Billy's intervention. Though similar, the concept of an ennobler of C with respect to E (where C is a cause of E) is not the same as that of a reason why C is a cause of E. The presence of oxygen is a reason why the striking caused the match to light. But the presence of oxygen is not an ennobler. It is false that without the oxygen the lighting of the match does not require the match to be struck. For without the oxygen the match would not light at all.[15] Still, it is plausible that every ennobler of C with respect to E is a reason why C is a cause of E. It is just that being an ennobler is not the only way to be a reason of this kind.

Though it's not relevant to my current line of thought, I have another idea about ennoblers that I want to share. An ennobler, I said, is a fact "without which the cause would not be required for the effect." Yablo actually wrote that ennoblers are "conditions such that the cause ceases to be required for the effect, if you imagine them away" (p. 98). But he soon translates this talk of "imagining away" into counterfactual terms: if C is a cause of E, an ennobler of C with respect to E is a fact G such

---

[15] In addition to ennoblers, Yablo also mentions what he calls "enablers"; the presence of oxygen is an enabler. I will say more about Yablo's notion of an enabler in Appendix B.

that, had C not occurred, but G still obtained, E would not have occurred (p. 99). Is this a *definition*? It shouldn't be; given the repeated failures of counterfactual theories of this and that in philosophy, we do best to regard this as a counterfactual *test* for ennobling, not a definition. But if the counterfactual is only a useful guide to which things are and are not ennoblers, we still need to know what it is to *be* an ennobler. Here's an idea: to be an ennobler of C with respect to E is to be a *reason why* E required C.

The witness to the truth of HIGH↛LOW I've focused on so far is also a witness to the truth of NON-TRANSITIVITY, but I do not think that every witness to the truth of the first is a witness to the truth of both. In fact I think that FIRST ROCK DIALOGUE, the dialogue about the dropped rock, contains a witness to the truth of HIGH↛LOW that is not a witness to the truth of NON-TRANSITIVITY. The fact that it is a law that $s = \sqrt{2dg}$ is a reason why ⟨the fact that the rock was dropped from one meter up is a reason why the rock hit at 4.4 m/s⟩. But (I hold that) the law is *not* a (first-level) reason why the rock hit at 4.4 m/s. Here is another high-level reason that fails to be a low-level reason. However, the example does not bear on the truth of NON-TRANSITIVITY—it does not bear on whether "reason why" is transitive—since the fact that it is a law that $s = \sqrt{2dg}$ is not a reason why the rock was dropped from one meter up.

If Carl Hempel had gone in for reasons-why talk, I think he would have disagreed; he would have rejected the claim that this example witnesses the truth of HIGH↛LOW. Hempel's DN model, again, says that an answer to the question why Q is a sound argument for the conclusion that Q that essentially contains a law as a premise. The most natural way to convert the DN model into a theory of reasons why has it say that each premise in a DN argument for the conclusion that Q is a reason why Q. Then the model entails that whenever there is an answer to the question why Q, some law is among the reasons why Q.

I think that it is a mistake to require, as this interpretation of the DN model does, every fact to have a law of nature among the reasons why it obtains. To think that the law that $s = \sqrt{2dg}$ is a reason why the rock hit at 4.4 m/s is to confuse higher-level reasons with first-level reasons. I think that Hempel was subject to this confusion, and inaugurated a long tradition of confusing the two levels of reasons. It is worth pausing to look at this tradition.

4.3 The History of a Mistake

Why did Hempel think that an answer to a why-question needed to cite a law of nature? One of Hempel's paradigm examples of a scientific explanation is the derivation of the state of a system at one time from its state at an earlier time and deterministic laws (see "Aspects," p. 351). The rock-dropping example is an over-simplified example of this kind.[16] The DN model can be made to look especially good if we imagine the dialogue about the dropped rock going differently, as in

SECOND ROCK DIALOGUE:
A: Why did the rock hit the ground at a speed of 4.4 m/s?
B: It landed at a speed of 4.4 m/s because it was dropped from 1 meter, and Newton's theory of gravitation entails that, for short falls, the impact speed s is related to the distance fallen d by the equation $s = \sqrt{2dg}$, where g is the gravitational acceleration near the surface of the earth. And of course $\sqrt{2 \cdot 1 \cdot 9.8} \approx 4.4$.

In FIRST ROCK DIALOGUE the law that $s = \sqrt{2dg}$ appeared in B's response to a vertical follow-up why-question, but here it appears in B's response to the original why-question. And what B says here looks like just as good a response to A's (first) question. In fact it looks better. For in FIRST ROCK DIALOGUE B's first response left A befuddled; this time, A will probably be satisfied. If SECOND ROCK DIALOGUE were our only data point, we would do well to guess that the law that $s = \sqrt{2dg}$ is part of B's answer to A's question, and that therefore this law is a (first-level) reason why the rock hit the ground at 4.4 m/s.

This is not, however, the role that I think the law is playing in B's response to A in SECOND ROCK DIALOGUE. The law, while certainly part of B's response, is not part of B's answer.

B's response to A is certainly a good response; but, as I noted in section 3.1, not every good response to a question is an answer to that question. In my example in section 3.1 someone replies "he's sick" to the question whether Connor is coming to the party. This response is a good one, even though it is not an answer, because it conversationally implies

---

[16] Oversimplified because it does not involve the solving of differential equations, but instead invokes an equation that directly connects the initial state (height) to the final state (impact velocity). There are differential equations lying behind the example, but they have already been solved.

an answer. But this is not exactly what I think is going on in SECOND ROCK DIALOGUE. My view is that B's response in SECOND ROCK DIALOGUE has two parts, an answer to A's question—it offers the fact that the rock was dropped from 1 meter as a reason why it hit the ground at 4.4 m/s—and also some other information, a law, that is not an answer, or part of an answer. B did not, by providing this extra information, conversationally imply an answer to A's question. So what is this extra information doing there? How is its presence in B's response compatible with that response being a good one?

Sometimes it is fine, in fact a good idea, the kind of thing a cooperative conversational partner would do, for one to give more information than is strictly required to answer the question one has been asked. If someone asks a question, we can sometimes (usually?) assume that she does not just want to know the answer to the question she explicitly asked but also wants to know the answer to other relevant questions. If you ask "Is Connor coming to the party?", and I know that you think that Connor *should* come to the party, then I know that if Connor is not coming, you will want to know why. In this case the response "He's sick" does not just conversationally imply an answer to your question; it is also itself an answer to a relevant—though unasked—follow-up question.[17]

That is what I think is going on in SECOND ROCK DIALOGUE. Person A asks why the rock hit at a speed of 4.4 m/s, and B knows that the answer to this question (namely, because it was dropped from 1 meter) will immediately raise the question of why this answer is the answer. Since B wants to be cooperative, he includes the answer to this natural—but unasked—follow-up in his response to the initial question. That is how the law that $s = \sqrt{2dg}$ can appear in a response to the question of why the rock hit at a speed of 4.4 m/s without being a reason why the rock hit at a speed of 4.4 m/s.

It is true: I have asserted, and have been defending, the claim that the law that $s = \sqrt{2dg}$ is not a reason why the rock hit at a speed of 4.4 m/s, but I have not given any direct argument for this conclusion. I have not deduced that the law is not a reason from an independently

[17] Grice gets close to the idea that a good conversational partner will sometimes respond to questions with information that goes beyond the answer, when the speaker can anticipate that that information is relevant to other questions the listener wants to know the answer to ("Logic and Conversation," p. 38).

well-motivated theory of reasons why. That it is not a reason why of course follows from theory (T1), but my goal here is to defend (T1), so I can't appeal to it as a premise. My argument has had to be indirect. I have tried to show that some data, namely the fact that B's response in SECOND ROCK DIALOGUE is a good one, that appears to establish that the law is a first-level reason, can be made consistent with my theory. To those who think, on the basis of dialogues like SECOND ROCK DIALOGUE, that laws are sometimes reasons why events occur, I ask: are you sure this is not a level confusion? Are you sure that the law is really part of the answer to the question why some event occurred, instead of appearing next to the answer, playing the role of an answer to an unasked but anticipated vertical follow-up why-question?

The law that $s = \sqrt{2dg}$ is certainly a higher-level reason why; it is certainly a reason why ⟨the fact that the rock was dropped from 1 meter is a reason why it hit at a speed of 4.4 m/s⟩. Why isn't this enough? I generally don't like talking about who bears the burden of argument, but in this case I do think that the burden is on those who think that something that is a higher-level reason is also a lower-level reason. There are two facts: the fact that ⟨the fact that the rock was dropped from 1 meter is a reason why it hit at a speed of 4.4 m/s⟩; and the fact that the rock hit at a speed of 4.4 m/s. The law that $s = \sqrt{2dg}$ is a reason why the first fact obtains. Why think that it is also a reason why some other, distinct, fact, the second fact, obtains?

Someone might reply to all this that they now have the distinction between first- and second-level reasons firmly in mind, and even still, "That $s = \sqrt{2dg}$ is a reason why the rock hit the ground at 4.4 m/s" strikes them as true. This sentence does not strike me as true; I think it is false; but I have no further arguments to offer to persuade such a person.

Michael Scriven said something about the relation between laws and answers to why-questions back in 1959 that resembles my view.[18] Hempel's summary of, and response to, Scriven's idea, is instructive:

> [Scriven] argues that in so far as laws are relevant to an expla-
> nation, they will usually function as "role-justifying grounds" for
> it.... Explanations might then take the form 'q because p,' where
> the 'p'-clause mentions particular facts but no laws ... The citation

---

[18] In "Truisms as the Grounds for Historical Explanation"; see also his "Explanations, Predictions, and Laws."

of laws is appropriate, according to Scriven, not in response to the question 'Why q?', which 'q because p' serves to answer, but rather in response to the quite different question as to the grounds on which the facts mentioned in the 'p'-clause may be claimed to explain the facts referred to in the 'q'-clause. To include the relevant laws in the statement of the explanation itself would be, according to Scriven, to confound the statement of an explanation with a statement of its grounds.   ("Aspects," p. 359)

That is Scriven's view. It is not quite mine; as I understand him, Scriven holds that one cites laws to provide *evidence* that "Q because R" is true, while I hold that one cites laws as *reasons why* "Q because R" is true. Our views are not that far apart, though.[19] In most cases, the reasons why some proposition is true can serve as evidence for the fact that it is true. It is just that the converse is false: not all evidence for the truth of X consists in reasons why X is true. (That my math professor believes that $2 + 2 = 4$ is evidence that $2 + 2 = 4$ but certainly not a reason why $2 + 2 = 4$.) Hempel rejected Scriven's view:

Now it is quite true that in ordinary discourse and also in scientific contexts, a question of the form 'Why did such-and-such an event happen?' is often answered by a because-statement that cites only

---

[19] In 2004 Jonathan Dancy also came close to my claim that laws are higher-level reasons why but not first-level reasons why, in his book *Ethics Without Principles*: "A full causal explanation of an event might be thought of as one that specifies a sufficient set of events as causes. Now what about the laws governing the whole transaction? It looks as if they are not a proper part of this explanation at all, because it is a specification of a sufficient set of events, and no law is an event. The role of the laws lies elsewhere. They are the conditions required for those events to necessitate this one, and thereby they stand as enabling conditions for the explanation rather than as a proper part of it" (p. 46). Dancy says that no law is part of a "causal explanation," the laws are just "enabling conditions"; I say that no law is a reason why any event happened, the laws are just reasons why some other events are reasons why it happened. These appear to be different claims. For by "causal explanation" Dancy seems to mean "answer to the question why E happened consisting in a list of at least some of the causes of E." The fact that no law is part of a causal explanation, so understood, leaves it completely open whether any law is a reason why E happened.

Dancy's goal in chapter 3 of his book is to defend the idea that X can enable Y to play some role without X itself playing that role. This is clearly close to my claim HIGH$\nrightarrow$Low, that A can be a reason why ⟨B is a reason why C⟩ without A being a reason why C. But if we just go by the ordinary meaning of "enabler" in English, the idea of an enabler seems different from the idea of a higher-level reason why. Suppose I am chosen by lottery to be "team runner"; the condition of my legs enables me to play that role, but it does not seem to be a reason why I play that role.

certain particular facts—even in cases where the relevant laws could be stated. The explanation statement 'The ice cube melted because it was floating in water at room temperature' is an example. But as this example equally illustrates, an explanation as ordinarily formulated will often mention only some of a larger set of particular facts which jointly could explain the occurrence in question. It will forego mention of other factors, which are taken for granted, such as that the water as well as the surrounding air remained approximately at room temperature for an adequate time. Hence, in order to justify attributing an explanatory role to the facts as actually specified, one would have to cite here not only certain laws, but also the relevant particulars that had not been explicitly mentioned among the explanatory facts. Thus it is not clear why only laws should be singled out for the function of role-justification. And if statements of particular fact were equally allowed to serve as role-justifying grounds in explanations, then the distinction between explanatory facts and role-justifying grounds would become obscure and arbitrary.

(p. 359)

It was when I read this passage that I really became convinced of Hempel's greatness. These are deep and interesting challenges to Scriven's view that are easily converted into challenges to mine. We just need to change the passage from a complaint about the notion of role-justifying grounds to a complaint about higher-level reasons why: "if laws are higher-level reasons why ⟨these causes of E are reasons why E happened⟩, then other causes, not mentioned in the answer, are also higher-level reasons why ⟨these causes of E are reasons why E happened⟩; but then the distinction between higher-level and lower-level reasons why becomes obscure and arbitrary."

Is this right? *Is* the distinction obscure and arbitrary? I don't think so. Hempel's ice cube example is weird,[20] so switch back to the match-lighting example. Suppose I say that one reason why the match lit is that I struck it. I hold that the fact that oxygen was present is a reason why ⟨the striking is a reason why it lit⟩, and that the laws of chemistry

---

[20] I have had trouble replicating his experiment. When I put an ice cube in a cup of water, the water in the cup does not remain approximately at room temperature while the ice melts. Usually my goal, in fact, is to get the water to a noticeably lower temperature. Maybe Hempel's ice was floating in a huge bowl of water. (Thanks here to Jon Shaheen.)

are also reasons why ⟨the striking is a reason why it lit⟩. I do not think that the laws should be "singled out" for the role of high-level reasons. I agree with the starting-point of the Hempelian complaint. I also hold that the first second-level reason, the presence of oxygen, is also a first-level reason, and that the other higher-level reasons, the laws, are not. But I do not think that this makes the distinction between first- and second-level reasons why "obscure and arbitrary." I do not see what is obscure about the thesis that while both events and laws can be second-level reasons, only events can be first-level reasons why E happened. As for the charge of arbitrariness, the idea must be that it is arbitrary to allow both events and laws to be second-level reasons, but not to allow both to be first-level reasons why some event E happened. But first-level reasons (the ones we're talking about here) are reasons why some event happened; second-level reasons are not. Given that reasons at different levels have different sorts of "objects," what is arbitrary about denying that laws can be first-level reasons?

Did Hempel have any other reason for holding that laws must appear in answers to why-questions than the naturalness of including laws in responses to why-questions, as in SECOND ROCK DIALOGUE? Look at what he says when he argues that "causal explanations" fit the DN model:

> [a] causal explanation implicitly claims that there are general laws—let us say, Ll, L2, . . . , Lr—in virtue of which the occurrence of the causal antecedents mentioned in Cl, C2, . . . , Ck is a sufficient condition for the occurrence of the explanandum event. This relation between causal factors and effect is reflected in our schema (D-N): causal explanation is, at least implicitly, deductive-nomological.
>
> (p. 349)

"In virtue of" is today taken to be interdefinable with "ground": fact A obtains in virtue of fact B iff fact B grounds fact A. So, taken out of historical context, it is natural to hear Hempel's claim that "there are general laws . . . in virtue of which the occurrence of the causal antecedents . . . is a sufficient condition for [the effect]" as entailing that, whenever C is a cause of E, there are some laws that ground the fact that C is a cause of E. Since I think that grounds are reasons why, I hold that it follows from all this that some laws are reasons why C is a cause of E. Presumably it follows from this that those laws are reasons why ⟨the occurrence of C is a reason why E happened⟩. If, then, one rejected HIGH ↛ LOW, if one held that higher-level reasons why are always also first-level reasons why, one would conclude that whenever a cause of E is a reason why E happened, there are also some laws that are reasons why E happened—namely, the laws in virtue of which that cause is a cause. And this is very close to the idea, enshrined in the DN model, that answers to why-questions must cite laws.

But this line of argument fails, because HIGH $\not\to$ LOW is true. Higher-level reasons are not automatically first-level reasons, and (on my view) are certainly not when the higher-level reasons are laws.

I do not wish to say that Hempel would recognize this as an argument he meant to give, since my guess is that he would reject the intelligibility of "grounding." Still, this may be a case where Hempel was relying in his thinking on the notion of grounding without realizing it. In that case this argument may have played a role in maintaining his belief in the DN model, even if he would have explicitly rejected the intelligibility of one of the notions the argument employs.

The space of theories of answers to why-questions is a minefield where each mine is one of the standardly accepted examples of an answer (or non-answer) to a why-question. The flagpole . . . the eclipse . . . the barometer . . . the ink-bottle: I knock an ink bottle with my knee, thereby spilling ink on the carpet. Why is the carpet stained? Because I knocked the ink bottle over with my knee. This answer contains no laws.[21] One

---

[21] The example is Scriven's ("Truisms," p. 456). Hempel had a response, but it didn't work. He said that when someone answers the question why E happened by citing a cause of E but not any laws, he is presupposing, or implying, the existence of some laws that connect C and E (see "Aspects," p. 347 and p. 349). It is interesting to note that Hempel shifts here from the claim that causal claims are true "in virtue of " certain laws to the claim that such claims "presuppose" or "imply" certain laws. The shift is from metaphysics to pragmatics. Someone who asserts a causal claim presupposes that certain laws are true iff they believe, and believe their audience believes, that those laws are true; they (conversationally) imply that certain laws are true iff, roughly, the hypothesis that they believe that those laws are true is needed to make their utterance consistent with the assumption that they are a good conversational partner. (Standard references on presupposition and implicature are Stalnaker, "Pragmatic Presuppositions," and Grice, "Logic and Conversation.") Either way the idea is that one need not mention a law in an answer to a why-question as long as it is common ground in the conversation which law is needed to complete the answer, or as long as the speaker has communicated the law to his audience in some other way. This attempt to use presupposition or implicature to defend the DN model is, I think, a failure. It is perfectly fine to say that the carpet is stained because I knocked the ink bottle without presupposing or implying the truth of any laws of nature. It is even perfectly fine to say this without presupposing or implying that *there are* any true laws of nature.

might enter this minefield at the DN model, see immediately that it is refuted by the ink bottle, move immediately to some theory that says that answers must say something about causes, come face to face with (what appears to be) an answer that cites no causes—an equilibrium explanation, for example—and then give up any hope of finding a simple unified theory. All of this is too fast.

The DN model is false. If you first see that it is false by encountering examples like Scriven's ink bottle, theory (T0) can look pretty good. But (T0) is harder to defend than some of its sympathizers realize, and also harder to refute than its opponents believe.

I have been working very hard to explain why SECOND ROCK DIALOGUE, which appears to confirm the DN model, and to disconfirm (T0) and (T1), does not do these things. Partisans of theories like (T0) have not appreciated how difficult it is to do this.

I should say that some partisans of theories like (T0) have theories that are more liberal than (T0), and (as I would put them) do allow laws to be reasons why events occur. Wesley Salmon came to believe that explanations of events must cite causes. But examples in which a law appears to be part of an answer to the question why some event occurred are not threats to his theory, because his "causal-mechanical" theory of explanation allows answers to why-questions to cite "causal regularities" in addition to causes. Salmon says that his theory is "as much a covering-law conception" of explanation as the DN model.[22] I think that Salmon's causal-mechanical theory, like the DN model, is false.

Similarly, Railton called his theory of explanations of chance events a causal theory, but he required explanations, "ideal" ones at least, to cite laws. I think that Railton's theory, like the DN model, is false.[23]

Scriven held that no law is a part of an answer to a question why some event occurred, that one invokes laws only to justify an answer. But then why does SECOND ROCK DIALOGUE, in which the law appears to be part of the answer, seem so natural? Answering a question, and justifying your answer, are different activities, and B seems just to be answering the question. No one has asked him to justify it. I have tried to address these concerns; Scriven is not even aware of them.

---

[22] See *Scientific Explanation and the Causal Structure of the World*, pp. 274 and 262.
[23] See "A Deductive-Nomological Model"; "Probability, Explanation, and Information."

David Lewis, whose theory is very close to (T0), was sensitive to the threat that examples like SECOND ROCK DIALOGUE pose, but he failed to counter it. At one point he said that a DN argument for the conclusion that some event E occurred that "looks explanatory" is no threat to his theory, as long as it "presents" causes of the event E ("Causal Explanation," pp. 234–5). But that's not enough. If a DN argument constitutes an answer to the question why E happened—as Lewis concedes it sometimes does— then *each part* of the argument is part of the answer. Lewis's view does *not* seem to be that producing a DN argument in response to "Why did E happen?" is like producing "John came to the party and the moon is full" in response to "Who came to the party?". He wasn't trying to say that only part of a DN argument can be an answer to a why-question; he seemed willing to accept that an entire DN argument can. How exactly is this compatible with his theory?

You might think that I am forgetting that the focus of his theory is on partial answers. Lewis held that a partial answer to the question why E happened is a proposition that says something about the causes of E. Lewis could say that the law in, for example, B's answer in SECOND ROCK DIALOGUE, says something about the causes of the rock's impact speed. This is a strategy he employs when discussing other examples: say that when a law appears in an answer to a why-question about an event, its role is to provide information about the event's causes.

But this strategy doesn't work. What does the fact that the impact speed s and fall distance d are related by the equation $s = \sqrt{2dg}$ say about the causes of the fact that the rock hit the ground at a speed of 4.4 m/s? The answer might seem obvious: it says that one of the causes (namely the drop height) is connected to the effect via this law. That is certainly a fact about E's causes. But to say this is to cheat. It is not a fact about E's causes in the relevant sense. According to Lewis, once again, a partial answer to the question why E happened must rule out some possible complete answer, and the complete answer specifies all the causes of E. The sense, therefore, in which a partial answer is a proposition about E's causes is this: a partial answer is a proposition about *what E's causes are*. ("E's causes," in "proposition about E's causes," is another concealed question.) A proposition that is about *how E's causes are related to it*, or about *what E's causes are like*, or about *what more fundamental events constitute E's causes*, not to mention a proposition about *how many words the shortest description of E's causes in English contains*, are not about E's causes in the relevant sense, and so do not constitute partial answers.[24] In SECOND ROCK DIALOGUE person B already listed one cause of the impact speed, namely the drop height. So adding that $s = \sqrt{2dg}$ to this fact is only a relevant thing to do, if Lewis's theory is correct, if it provides further information about what E's causes are. But it does not do that. The law does not appear to tell us anything about what E's other causes are. And even if it did, the fact that it did this was not B's reason for mentioning the law; B included the law in his response because of what it says about the relationship between impact speed and the drop height, not because it tells us anything about the impact speed's other causes.

Despite what Lewis says about examples like SECOND ROCK DIALOGUE, then, it seems to me that if every part of B's response to A's question in SECOND ROCK DIALOGUE is part of an *answer* to A's question—and Lewis does not deny this—then Lewis's theory is false.

The thing to say about SECOND ROCK DIALOGUE is that not everything B says is part of his answer. Then one must say something about how the rest of what he says can be relevant. I have said that the rest of what he says constitutes higher-level reasons, and have argued that providing such things is often a relevant thing to do.

Lewis did not confuse higher-level with first-level reasons why; he did not make the mistake the history of which I am chronicling. But neither did he distinguish between the two levels, and it was another flaw in his theory. The distinction is needed to defend theories (T0) and (T1) against examples like SECOND ROCK DIALOGUE.

In the 1970s Wesley Salmon proposed a "Statistical-Relevance" (SR) model of explanation.[25] It is (explicitly[26]) a theory of answers to why-questions of the form "Why is this thing X, which is an A, also a

---

[24] Railton ("Probability, Explanation, and Information," p. 246) was willing to say that a proposition about the length in English of a description of E's causes can constitute a partial answer to the question why E happened. This is crazy. Why was Railton willing to say this? Because he had a false theory of partial answers. He did not say that a proposition is a partial answer iff it provides some information about what the complete answer is; he said that a proposition is a partial answer iff it provides some information about what the "ideal explanatory text" is, where this is some set, or list, of sentences that expresses the complete answer.

[25] Salmon's summary statement of his view is on pp. 76–77 of *Statistical Explanation and Statistical Relevance*. I have rephrased it slightly in inessential ways. Salmon later revised and abandoned the SR model (see his *Scientific Explanation and the Causal Structure of the World*). My earlier mention of Salmon referred to his later views.

[26] That is, I did not have to translate Salmon's theory from the form "F explains E iff..."

B?" Roughly speaking, Salmon held that an answer provides a list of statistically relevant properties, and says which of these properties X has. In more detail, an answer includes all of the following facts:

- The probability that something is a B, given that it is an A and also a C1, is p1;
- The probability that something is a B, given that it is a A, and also a C2, is p2; [this list could go on, up to Cn; for simplicity I will assume that n=2];
- The set of properties {the property of being an A and a C1; the property of being an A and a C2} is a "homogeneous partition of the property of being an A with respect to the property of being a B." That is, every A is either a C1 or a C2, and no A is both (that's the partition part); and for any Y and Z that are A and also C1, the probability that Y is a B is the same as the probability that Z is a B, and similarly for C2 (that's the homogeneous part);
- p1 ≠ p2;
- X (which is an A) is a C1 (or a C2, whichever is true).

This theory is on its face pretty crazy. Why is Jones, who is a person who contracted strep throat, a person who recovered from strep throat within a week? Here's what Salmon thinks an answer should look like:

> STREP: The probability that someone with strep recovers within a week, given that he takes penicillin, is 90%; the probability that someone with strep recovers within a week, given that he does not take penicillin, is 30%;[27] {The property of being someone who has strep and takes penicillin; the property of being someone who has strep and doesn't take penicillin} is a homogeneous partition of the property of being someone who has strep with respect to the property of recovering within a week;[28] 90% ≠ 30%; and Jones took penicillin.

No one has ever given a speech like this in response to a question like "Why did Jones recover from strep within a week?" What we all say, instead, is simply "Because Jones took penicillin."

---

[27] I've made these numbers up.

[28] This is of course false; not everyone who has strep and takes penicillin has the same chance of recovery within a week. Some have a strain of strep that is resistant to penicillin. For the sake of having a simple example I am going to ignore this complication.

Salmon's view, I take it, is that all the other facts that appear in STREP are really part of the answer to the question (in my preferred terms, they really are reasons why Jones recovered in a week), but we often omit those other parts when we don't know them, or when we assume that our audience already knows them. I find this incredible. But I can also see why Salmon wants to require those facts to be in the answer. He doesn't have the distinction between first-level and second-level reasons why. He thinks that those other facts are relevant in some way to the question asked, and the only way he can make them relevant is to say that they are part of the answer, and therefore are reasons why Jones recovered within a week.

But once we recognize the distinction between levels it is no longer necessary to insist that, for example, the fact that taking penicillin made Jones's chance of recovering from strep 90% (when it would have been 30% had he not taken it) is a reason why Jones recovered in a week. We can say instead that the only first-level reason why Jones recovered is that he took penicillin. We can then say that the fact that his chance of recovery in a week depended on whether he took penicillin is a second-level reason, a reason why the fact that Jones took penicillin is a reason why he recovered.

A doctor might say something closer to what Salmon says the answer is, especially if talking to someone who is not a doctor: "Jones recovered in a week because he took penicillin, which makes recovery in a week much more likely." But we need not take the fact that doctors sometimes give responses like this as evidence that favors Salmon's theory. It's just like with the DN model. This doctor is providing both the answer to the question why Jones recovered in a week, and also the answer to the unasked follow-up question, why is the taking of the penicillin a reason why Jones recovered in a week?

If we divide the facts that Salmon says belong in answers to why-questions into the facts that really are first-level reasons, and those that are second-level reasons, in the way that I have done, Salmon's theory begins to look a lot like (T0), the theory that the reasons why some event E happened are its causes. For the fact that Jones took penicillin did not just raise the chance of his recovery, it caused his recovery. Moreover, it is at least initially plausible to hold that the fact that taking penicillin raised the chance of recovery is the reason why taking penicillin caused his recovery.

Salmon (even in 1971) was certainly sensitive to the idea that an answer to a why-question about an event can cite causes and shouldn't cite non-causes. Hempel had earlier advanced the theory that a statistical explanation of the occurrence of an event E is a strong inductive argument that has the proposition that E happened as its conclusion and contains at least one law as a premise (where a strong inductive argument has true premises that confer high probability on its conclusion).[29] Salmon objected that the following strong inductive argument is not an answer to the question why Smith recovered from his cold within a week:

- Smith took high doses of vitamin C.
- The probability that someone recovers from his cold within a week, given that he took high doses of vitamin C, is very high.
- Therefore, Smith recovered from his cold within a week.

The premises make the conclusion very likely, but Smith's taking vitamin C is not why he recovered. The problem, Salmon observed, is that people who *don't* take vitamin C are also very likely—in fact just as likely—to recover from their cold within a week. After making this observation about the statistics Salmon explicitly connected statistical relevance (in his sense) to causation: "Vitamin C is not efficacious, and that fact is reflected in the statistical irrelevance of administration of vitamin C to recovery from a cold within a week" (p. 47). Still, Salmon (in 1971) did not accept anything close to (T0). The SR model is formulated in terms of statistical relevance, not causation.

Salmon closed his essay with this thought:

> I should be inclined to harbor serious misgivings about the adequacy of my view of statistical explanation if the statistical analysis of causation cannot be carried through successfully, for the relation between causation and explanation seems extremely intimate. (p. 81)

So Salmon held that a collection of facts constitutes an answer to the question why E happened iff those facts display factors that are statistically relevant to the occurrence of E, and also display that those facts are statistically relevant; he also held, or hoped, that some analysis of causation in statistical terms, to which "C is a cause of E iff it is statistically relevant to the occurrence of E" may be a first approximation, is correct. From my point of view, all the pieces are in the wrong place. Maybe some statistical analysis of causation is correct (though existing ones face challenges[30]). But if one is drawn to statistical analyses of causation, it is a mistake

---

[29] See "Aspects," section 3.3.   [30] See Hitchcock, "Probabilistic Causation."

to have a theory of answers to why-questions that mentions statistical relevance but not causation. The right thing to say is that statistical facts are higher-level reasons why causes are causes,[31] and therefore are higher-level reasons why causes are reasons why their effects obtain. It is a level confusion to then also regard the statistical facts as reasons why those effects obtain.

This section is already too long. So fast-forward thirty years: James Woodward publishes *Making Things Happen,* which defends a "manipulationist" theory of explanation. Woodward's theory, or at least the framework his theory inhabits, has been hugely influential.[32] Nevertheless, the theory is—I believe—false, and rests on a confusion of levels of reasons why in much the same way the DN model did forty years earlier.

Woodward and Hitchcock's "provisional formulation" of the manipulationist theory, in their paper "Explanatory Generalizations, Part I," reads as follows:[33]

> An explanation involves two components, the explanans [facts that do the explaining, that answer the relevant why-question] and the explanandum [the fact about which we have asked why it obtains]. The explanandum is a true (or approximately true) proposition to the effect that some variable (the 'explanandum variable') takes on some particular value. The explanans [the answer] is a set of propositions, some which specify the actual (approximate) values of variables (explanans variables), and others which specify relationships between the explanans and explanandum variables. These relationships must satisfy two conditions: they must be true (or approximately so) of the actual values of the explanans and explanandum variables, and they must be invariant under interventions. (p. 6)

"Note," they continue, "the similarity in structure between this formulation and the formulation of Hempel's D-N theory of explanation":

---

[31] The first occurence of "causes" in "reasons why causes are causes" is to be read *de re* (I'm not discussing the reasons why some logical truth is true); the same goes throughout this book for occurrences of "the cause" in contexts like "why the cause is a cause" and "the fact that the cause is a cause."

[32] As of this writing, *Making Things Happen* has almost 2000 citations on Google Scholar. Hempel's *Aspects of Scientific Explanation*, the founding document of post-positivist theorizing about explanation, has ("only") 5800, despite being five times older.

[33] A very similar summary appears on p. 203 of *Making Things Happen*.

The statement specifying the value of the explanandum variable is analogous to Hempel's explanandum proposition; the statements specifying the values of the explanans variables are analogous to Hempel's initial conditions; and the invariant generalizations figuring in our explanans are analogous to Hempel's laws. This similarity of structure will help to bring the essential differences between the two accounts into sharper focus.   (p. 6)

Woodward and Hitchcock's statement of their theory is thick with the terminology of variables and values of variables; read as a theory of answers to why-questions, it is a theory of what it takes to answer the question why some variable V took on some value x, not what it takes to answer the question why some event E occurred (the kind of question I have focused on). Throughout their work Woodward and Hitchcock make a good case that variable-speak is more powerful than event-speak; it allows us to make distinctions explicit that are often hidden when we use the language of events, distinctions that are important for answering why-questions. But what I want to say about their theory does not require this extra power, so I will stick with the language of events.[34] First let's see the summary of the manipulationist theory rewritten to (i) speak of events, and (ii) be a theory of answers to why-questions, rather than a theory of "explanation":[35]

An answer to the question why some event E happened is a set of propositions, some which state (truly) that certain events C1, . . . ,Cn occurred, and others which specify relationships between E and C1, . . . ,Cn. These relationships must satisfy two conditions: they must be true (or approximately so) of these events, and they must be invariant under interventions.

The events C1, . . . ,Cn are causes of E.[36] Stating that they occurred is not enough; the manipulationist theory requires an answer to the question

---

[34] I will translate my claims into variable-speak in the footnotes.

[35] Officially, the theory is a theory of "causal explanation," not a theory of explanation-in-general. I will leave this qualification tacit.

[36] Or: for each of the "explanans" variables, the fact that it takes on a specified value is a cause of the fact that the "explanandum" variable takes on its specified value. (Woodward and Hitchcock do not come right out and say this in the passage quoted, but it is clear from the rest of the theory.)

why E happened to do more. It must also articulate a true relationship between the causes and the effect that is "invariant under interventions."

By now it will come as no surprise when I say that requiring "an answer" to the question why E happened to contain an invariant generalization is ambiguous; there are several types of answers. I am going to interpret the theory to say that, for every event, some invariant generalization relating that event to some of its causes is a reason why that event happened.

But what it is for a relationship among events to be invariant under interventions? Woodward spends dozens of pages of his book making clear what is meant by "intervention" in this theory (see chapter 3 of *Making Things Happen*), but I must be brief. It is easiest to look at an example. Suzy throws a rock, which breaks a window; no one else was around. Why did the window break? Manipulationism says that these are the reasons why:

(6)   Suzy threw a rock at the window; and

(7)   The window broke if and only if Suzy threw a rock at the window.

(6) is the "cause-stating" part of the answer; (7) is the invariant generalization. It states a true relationship between the throw and the break, namely that either they both happened, or neither happened. Why is (7) invariant? Roughly speaking, for (7) to be invariant under interventions it must be the case that (7) would still have been true if Suzy had not thrown the rock. More carefully, it must be that (7) would still have been true if (i) Suzy had not thrown the rock, and (ii) whatever would have made it the case that Suzy did not throw the rock would have accomplished this without influencing whether the window broke in any way other than by influencing whether Suzy threw. Condition (ii) is not an easy condition to wrap your mind around when stated informally like this, but it is not hard to think of particular ways it could have come about that Suzy did not throw that satisfy (ii). Suppose Suzy had a change of heart. Or she was distracted for a moment by a passing bird, and in the moment realized she had something better to do. Neither the change of heart nor the passing bird has any influence over the window other than by influencing whether Suzy threw.[37]

---

[37] In terms of variables the example is this: let S have value 1 if Suzy throws, 0 otherwise; let W have value 1 if the window breaks, 0 otherwise. Then the manipulationist answer to the question why W=1 is that S=1—that's the variable-value stating part of the answer—and

We saw Hitchcock and Woodward explicitly draw a connection between their requirement that answers to why-questions contain invariant generalizations and Hempel's requirement that answers to why-questions contain laws. Although the actors are different, I think that the play is the same.

In this simple case, the fact that (7) is an invariant generalization amounts, more or less, to the truth of the counterfactual that if Suzy hadn't thrown (as a result of an intervention), the window would not have broken. I can see why one might think that the truth of this counterfactual has something to do with the question why the window broke. But I don't think it is part of the answer. I don't think it is a reason why the window broke. The counterfactual, instead, looks like a good thing to say in response to the question, "Why is the fact that Suzy threw a reason why the window broke?" The truth of the counterfactual, I hold, is a higher-level reason why, a reason why ⟨the throw is a reason why the window broke⟩. It is a level confusion to say that the counterfactual is also a reason why the window broke.[38]

Here is the shortest path from the DN model to Woodward's theory. Think again about SECOND ROCK DIALOGUE:

A: Why did the rock hit the ground at a speed of 4.4 m/s?

B: It landed at a speed of 4.4 m/s because I dropped it from 1 meter, and Newton's theory of gravitation entails that, for short falls, the impact speed s is related to the distance fallen d by the equation $s = \sqrt{2dg}$, where g is the gravitational acceleration near the surface of the earth. And of course $\sqrt{2 \cdot 1 \cdot 9.8} \approx 4.4$.

The fact that s and d are related by $s = \sqrt{2dg}$ seems like an important part of B's answer. Hempel thought: that's because that relationship is a law—explanations must cite laws! However, while it is true that the fact that $s = \sqrt{2dg}$ is a law, it is also true that it is an invariant generalization. Had the drop height been different as a result of an intervention (say I moved

S=W—that's the invariant generalization. It is invariant because it still would have been true if S had had value 0 as a result of an intervention.

[38] Really the truth of the counterfactual is a higher-level reason why only if some counterfactual theory of causation is right. Whether a counterfactual theory of causation is correct is contentious. But this is beside the point here. My claim is that, insofar as the truth of the counterfactual "Had Suzy not thrown, the window would not have broken" has anything to do with the question why the window broke, it is (merely) an answer to a vertical follow-up to the question, not part of the answer to the question itself.

my hand and dropped the rock from a different height), it still would have been true that $s = \sqrt{2dg}$. Not every law is an invariant generalization, not every invariant generalization is a law. Woodward thought that Hempel reached for the wrong conclusion and drew the other one: explanations must cite invariant generalizations. I think that both these philosophers went wrong right at the beginning. The equation $s = \sqrt{2dg}$ was not (I hold) part of B's answer at all. Recognize this, and the pressure to find a generalization about answers to why-questions that covers the inclusion of $s = \sqrt{2dg}$ in this case vanishes.

Hempel knew he needed to say something about answers to why-questions like "The carpet is stained because I knocked the ink bottle over with my knee," which only describe causes, and so don't state any laws of nature. The ink bottle example doesn't state any invariant generalizations either, and so also looks like a counterexample to the manipulationist theory. Woodward, like Hempel, was aware of these kinds of examples, and tried to make them compatible with his theory. I think he failed, but will leave my argument for Appendix A.

4.4 Return to Equilibrium Explanations

A long time ago I was talking about equilibrium explanations, and whether they are compatible with (T1). The problem was that Fisher presented a body of fact, Fisher's Information, in response to the question why humans are disposed to have sons and daughters in equal numbers, and many of the facts in Fisher's Information were not causes of the fact that humans are disposed to have sons and daughters in equal numbers. My initial response to the example was that we do not need to regard those facts as reasons why humans are disposed to have sons and daughters in equal numbers in order to see them as good things to include in a response to the question. I said that they could be good things to include even if they were (merely) "epistemic enablers": they were there to help those reading Fisher's response see that the fact that humans were never disposed to have only sons or only daughters is the only (relevant) reason why humans now are disposed to have sons and daughters in equal numbers.

With the distinction between levels of reasons why I can say more. The subjunctive conditionals that make up the bulk of Fisher's Information are all higher-level reasons why. They are reasons why ⟨the fact that humans were never disposed to have only sons or only daughters in the past is a reason why humans are now disposed to have sons and daughters in equal numbers).

Higher-level reasons why, of course, are well-suited to play the role of epistemic enablers. A very good way to help someone see that p is true is to present them with reasons why p is true.

The idea that it is natural for a response to a why-question to include second-order reasons why as well as first-order ones, to include not just an answer to the question asked but also answers to unasked vertical follow-up questions, is (to mix a metaphor) the key to defusing many apparent counterexamples to (T1), as we are about to see.

Appendix A: More on Manipulationism

Suppose lightning strikes in a field, causing the field to catch on fire. Then no matter what else is true (the reason for this qualification will emerge), the following is certainly a causal explanation of the fire:

LIGHTNING: The field caught on fire because it was struck by lightning.

Does Woodward's manipulationist theory count this as a causal explanation? I have found it exceedingly difficult to settle on an answer. I have spent far more time on this question than I would care to admit. Every time I thought I had pinned down what Woodward's theory says about LIGHTNING, it all slipped through my fingers like so much water.

It is clear that, on Woodward's theory, being a causal explanation is closely connected with in some sense exhibiting, or providing, counterfactual information. In Woodward's words:[39]

"[causal] explanation is a matter of exhibiting systematic patterns of counterfactual dependence."   (p. 191)

"the underlying or unifying idea in the notion of causal explanation is the idea that an explanation must answer a what-if-things-had-

---

[39]   All quotations in this appendix are from *Making Things Happen*.

been-different question, or exhibit information about a pattern of dependency."   (p. 201)

"The theory of causal explanation I have been sketching is thus one that ties explanatory import very closely to the provision of certain kinds of counterfactual information: it might fairly be described as a counterfactual theory of causal explanation."   (p. 196)[40]

Earlier, in section 3.3, I focused on a statement of Woodward's theory that required causal explanations to cite invariant generalizations. Is that statement equivalent to these ones, that require counterfactuals? Woodward seems to think so, since he moves back and forth between them without comment. I have my doubts, but will not take up the question. Since it is the statements that require counterfactuals that are relevant to what Woodward wants to say about (what he calls) "singular causal explanations," I will focus on them in this appendix.

The above statements of his view are somewhat vague. The last of them for example says that explanatory import is "closely tied" to counterfactual information; how closely are they tied? What kind of tie is this? To get some answers let us look at what Woodward says about "singular causal explanation."

At the beginning of section 5.8 of his book, which bears the title "Singular-Causal Explanation," Woodward writes that "[t]oken-causal claims imply various counterfactuals, and it is in virtue of conveying this counterfactual information that we should think of them as explanatory" (p. 210). He then discusses several sentences similar to

(A)   The lightning caused the fire.

He asks, "what can we say about [their] counterfactual import . . . or about what they tell us by way of answers to what-if-things-had-been-different questions" (p. 211)? Here is his answer (I have substituted my example for his):

Let us begin with the simplest sort of situation in which [(A)] might be used, a situation in which there is no causal overdetermination:

---

[40] If Woodward has a counterfactual theory of causal explanation, I guess I have a causal theory of causal explanation. This contrast makes me feel a bit like Howard Dean appealing, during the 2004 US presidential election, to the "democratic wing" of the Democratic party. (It does not please me to remember that Dean lost the race for the Democratic nomination.)

[no other cause of the fire besides the lightning, say an arsonist, is] operative or waiting in the wings. Along with most other commentators who favor counterfactual accounts of causation, I claim that in this sort of case, a singular causal claim (or explanation) of form c caused e implies the following counterfactual: if c had not occurred, then e would not have occurred. In particular, this counterfactual will hold even if c is an indeterministic cause of e. For convenience, I call this a "not-not" counterfactual. Thus, [(A)] (or strictly speaking, [(A)] in conjunction with the information that other causes of the fire are absent) implies the following not-not counterfactual:

[(CF)   If the lightning had not occurred, the fire would not have occurred.]

. . . We can think of this counterfactual information just as it stands as conveying information about the answer to what-if-things-had-been-different questions and as (at least minimally) explanatory for just this reason. . . . Thus, [(CF)] may be regarded as explaining [the fire] because it identifies a condition ([the lightning]) such that if this had been different (in particular, if it had been absent), this explanandum phenomenon [event being explained] would have been different (in fact, would not have occurred). . . . As with other causal and explanatory claims, these counterfactuals should be understood as claims about what would happen if interventions were to occur; thus, [(CF)] should be interpreted as claiming that if the [lightning] had been caused not to occur as a result of an intervention, then the fire would not have occurred, and so on.   (p. 211)

What Woodward says here does not directly apply to LIGHTNING, since LIGHTNING is not of the form "c caused e." Amazingly, Woodward never discusses sentences like LIGHTNING, sentences of the form "Q because R" where what goes in for "R" expresses a cause of the fact expressed by what goes in for "Q." He never says anything about what it takes for them to be true, even in this section, which again is titled "Singular-Causal Explanation." This despite the fact that sentences like LIGHTNING are the most central paradigm cases of causal explanation.

Any theory of causal explanation worth its salt must say what it takes for LIGHTNING to be true. Since LIGHTNING is my focus, I will have to use what Woodward says about why (A) ("The lightning caused the fire") is explanatory to hazard some guesses about what he would say about why LIGHTNING is true. Since he thinks that (A) is explanatory in virtue of which counterfactuals it "implies," I will assume that LIGHTNING implies whatever counterfactuals (A) implies. (This assumption is probably false, but I don't really know how else to proceed.)

Of course LIGHTNING could be true even if there *were* an arsonist waiting in his car to light the field on fire if the lightning storm didn't do it for him. If there were such an arsonist then LIGHTNING would be true, but the counterfactual (CF) would be false. Woodward is of course aware of this. He would say that in this new scenario, (A) (and presumably LIGHTNING as well) counts as a causal explanation in virtue of implying the truth of a different counterfactual, namely

(CF-2) Had lightning not struck, and the arsonist (still) not acted, the field would not have caught on fire.[41]

At this point I will play my usual hand. To know what to think of Woodward's claims, I need to know what they look like when stated as claims about the reasons why the field caught on fire. So what do they look like? I don't know. Let's look at our options.

To set them out let us assume that LIGHTNING is used in a circumstance of the kind Woodward focuses on first: there are no backups present ready to cause the fire if the lightning fails. We can start with

Option 1: The fact that the counterfactual (CF) is true is a reason why the field caught on fire. The fact that lightning struck the field is *not* a reason why the field caught on fire. When one "cites" this cause by uttering the sentence in LIGHTNING, what one says is true only because one "implies" that the truth of (CF) is a reason why the field caught on fire.

Here are some claims about reasons why that I understand. Support for Option 1 as an interpretation of Woodward comes from the fact that, in

---

[41] For confirmation that Woodward would say this, see his discussion on p. 219. Like (CF), (CF-2) should be interpreted as an "interventionist" counterfactual. That is, it should be interpreted to be equivalent to "Had lightning not struck *as a result of an intervention,* and the arsonist (still) not acted *as a result of an intervention,* the field would not have caught on fire." This qualification applies throughout this appendix, but I will leave it tacit.

the long quotation above, he starts by saying that the causal claim (A) implies the counterfactual, and then focuses on arguing that the counterfactual is explanatory. Later in the book he also calls counterfactuals like (CF) "carriers for the explanatory content" of claims like (A) (p. 220).

I think that the claims in Option 1 are false. First, it is absurd to say that the lightning is not a reason why the field caught on fire. Second, since Option 1 says that LIGHTNING is, nevertheless, true, it conflicts with the very plausible principle that a sentence of the form "Q because R" is true only if "one reason why Q is that R" is true. A third problem concerns what Option 1 says *are* the truth-conditions for LIGHTNING. It is hard to see how the *truth* of LIGHTNING could depend on what is *implied* by someone who utters it. And fourth, even if we set the third point aside, it is false that whenever someone truly utters the sentence in LIGHTNING in a situation with no backups, he implies that (CF) is a reason why the field caught on fire. The most natural interpretation of "implies" as it occurs in Option 1 is "conversationally implies." Now you can only conversationally imply something if you believe it. And someone could *be* in a situation where no backups are prepared to start the fire without *believing* himself to be in such a situation, and therefore without believing (CF), and (so) without believing that (CF) is a reason why the field caught fire. He might have no idea whether any backups are present. Even in this sadly impoverished epistemic state, he could know that LIGHTNING was true, he could truly utter LIGHTNING to answer the question why the field caught on fire. But by uttering LIGHTNING he could not conversationally imply that the truth of (CF) is a reason why the field caught on fire.

Maybe, though, all of Woodward's talk of "implying" and "exhibiting" counterfactuals is not meant to be understood in terms of conversational implicature. Maybe it is meant to be understood in terms of one proposition entailing another. Which propositions? The claim must be that the proposition that the lightning caused the fire entails the counterfactual (CF). Of course this cannot quite be right. That the lightning caused the fire does not entail (CF) all by itself; the former could be true, and the latter false, if there were backups present. But it is plausible that this claim about causation does entail (CF) given the right background assumptions. The obvious assumptions to make are: an appropriate "causal model" of the situation (which will include the fact that no backups are present), and Woodward's counterfactual theory of (token) causation. Summarizing Woodward's theory of causation would take a while, but fortunately I think we can make do with a rough understanding. His is a "conditional dependence" theory of causation.[42] In outline, it says that C is a cause of E iff, holding fixed certain facts, if C hadn't happened, E wouldn't have happened. The hard part of a theory like this is stating the criteria for what facts may be held fixed in testing for causation. We are not going to need these details (anyway, (CF) does not involve holding any facts fixed). This line of thought leads us to another interpretation, namely

> Option 2: The fact that the counterfactual (CF) is true is a reason why the field caught on fire. The fact that lightning struck the field is *not* a reason why the field caught on fire. When one "cites" this cause by uttering the sentence in LIGHTNING, what one says is true only because LIGHTNING, together with an appropriate causal model of the situation, and Woodward's theory of causation, entails (CF).

I think the claims in Option 2 are also false. It is still absurd to say that the lightning is not a reason why the field caught fire. Option 2 is also deeply in tension with one of Woodward's core commitments. This commitment first comes out when Woodward discusses the DN model:

> Hempel's overall strategy of trying to understand how explanations like [LIGHTNING] work [is] by treating them as devices for conveying information, but in a "partial" or "incomplete" way, about underlying "ideal" explanations of a prima facie quite different form that are at least partly epistemically hidden from those who use the original, nonideal explanation. (p. 159)

The reason why part of the "ideal DN explanation" that is associated with LIGHTNING is "epistemically hidden" from those who use LIGHTNING is that people who use LIGHTNING typically do not know of any laws that could be added to the clause following "because" to form a complete DN argument. Woodward thinks that this strategy is "deeply problematic" (p. 159). But Woodward's own theory, as interpreted by Option 2, uses this same strategy. It treats LIGHTNING as a device for entailing the

---

[42] The theory makes heavy use of "structural equations"; theories of this kind draw heavily on *Causality* by Pearl and *Causation, Prediction, and Search* by Spirtes et al. Woodward presents the theory he prefers in chapters 2 and 3 of *Making Things Happen*, and acknowledges that it is not original to him. In addition to Pearl's book he credits Halpern and Pearl's *Causes and Explanations* and Hitchcock's "Intransitivity" paper.

counterfactual (CF), which is the real reason why the field caught on fire. But someone may say LIGHTNING, and say something true, and be perfectly within his rights to say it, without knowing that it, together with an appropriate causal model, and so on, entails (CF) (and, for what it's worth, also without knowing that (CF) is true). For, again, someone may say LIGHTNING and say something true without knowing whether any backups are present.

Woodward explicitly addresses the question whether his theory does a better job avoiding the problem of epistemic hiddenness than the DN model does. Here is what he says:

> singular-causal explanations wear the source of their explanatory efficacy on their face: they explain not because they tacitly invoke a "hidden" law or statement of sufficient (or probabilifying) conditions, but because they identify conditions such that changes in these conditions would make a difference for whether the explanandum phenomenon [event being explained] or some specified alternatives would ensue. The information about such conditions and the counterfactuals associated with them are epistemically accessible and nonhidden.   (p. 217)

This just is not true. When a backup arsonist is present, LIGHTNING is true, but it does not "identify conditions such that changes in these conditions would make a difference for whether the event being explained would ensue." The only condition that LIGHTNING identifies is the lightning; and changing whether the lightning occurred would make no difference to whether the field caught on fire. Of course it is true that *there are* conditions such that changes in them would make a difference; changing both whether the lightning struck, and whether the arsonist is disposed to act if no lightning strikes, would make a difference. It is also true that, if a backup arsonist is present, and Woodward's theory of causation is right, then the truth of (CF-2), "If lightning hadn't struck, and the arsonist had (still) not acted, then the field would not have caught on fire," is sufficient for "The lightning caused the fire" to be true. But it is false that it is "epistemically accessible and nonhidden" which of (CF) and (CF-2) is true to anyone who is in a position to truly assert LIGHTNING, or to their audience.

Both Option 1 and Option 2 take pretty seriously Woodward's use of the word "imply," when he says that causal claims are explanatory because they "imply" counterfactuals. I can think of other interpretations that focus less on his use of "imply," and focus more on his use of words like "exhibit" or "provide" (forms of these words appear in the three quotations from Woodward at the beginning of this appendix). But these other interpretations are going to founder on the same hard rock that sinks Options 1 and 2. The root problem is that someone can truly assert LIGHTNING, even when no backups are present, without knowing that the counterfactual (CF) is true. It could even become quite common for this to happen—just imagine a town known to have a large population of arsonists who prefer to hide their fire-lighting activities under the cover of electric storms. Since someone can truly assert LIGHTNING without knowing that (CF) is true, that counterfactual is not always "epistemically accessible and nonhidden" when LIGHTNING is true (and appropriately asserted). Let someone to whom (CF) is epistemically hidden assert LIGHTNING; as long as there are no backups, he has said something true, but just as he has not conversationally implied that (CF) is true, he has not exhibited its truth, or provided it, or showed its truth to his audience,[43] or put his audience in a position to figure out on their own that it's true.

Options 1 and 2 contain the claim that the lightning is not a reason why the field caught on fire. I objected to them for containing this claim, but I hope it's clear by now that this is not my real problem with them. My real problem with them is the claim that (CF) is (also) a reason why the field caught on fire, and the claim that a use of the sentence in LIGHTNING is, in some sense, correct only if it somehow gets this counterfactual information across.

There is a final interpretation of Woodward's view I want to mention, inspired by how he starts his discussion of "singular causal explanation." Here is the quotation again:

> [t]oken-causal claims imply various counterfactuals, and it is in virtue of conveying this counterfactual information that we should think of them as explanatory.
>
> (p. 210)

I hold that if "H in virtue of G" is true, so is "G is a reason why H." If we replace the "in virtue of" language in this quotation with reason-why

---

[43] Woodward sometimes says that an explanation works by "showing" that certain counterfactuals are true; see, for instance, his discussion of his example (5.3.4) on p. 199.

language, and make some other adjustments (including ignoring the bit about "conveying"), we get

> Option 3: The reason why LIGHTNING is true in a scenario with no backup arsonists, the reason, that is, why it is true that the field caught on fire because it was struck by lightning, is that the counterfactual (CF) is true.

This is something I could go along with.[44] Option 3 does not require someone who gives LIGHTNING as an answer to the question why the field caught on fire to know, or believe, the counterfactual (CF), or to in any sense convey that counterfactual to his audience. It does not require this because there just is no general requirement that someone who knows the answer to a why-question, and gives that answer when asked that why-question, must know why that answer is the answer (must know why the reasons he gave are reasons). Still, although Option 3 is suggested by some of Woodward's wording, I don't really think that it is his view.

Appendix B: A Fully General Theory of Reasons Why?

Whatever goes in for "Q," whether it expresses a fact that corresponds to an event or not, every cause and every ground of the fact that Q is a reason why Q.[45] But not the other way around; (T2) from chapter 3, which asserts the other direction, is false. What counterexamples have we seen?

There were what Yablo calls ennoblers: Suzy's throw ennobles Billy's sticking out of his mitt from an event that preceded the window's survival into something that was required for its survival, and thereby into a cause of that survival. By being an ennobler, Suzy's throw is a reason why ⟨Billy's sticking out his mitt was (causally) required for the window's survival⟩, without being a cause or ground of this fact.

---

[44] Though I don't think that Option 3 is the full story. I think that one reason why ⟨the field caught on fire because it was struck by lightning⟩ is that the lightning caused the fire, and that the counterfactual (CF) is a reason why this cause is a reason only because it is a reason why this cause is a cause. I'll have more to say about the relationship between the reasons why something is a cause and the reasons why it is a reason in section 5.2.

[45] Of course when what goes in for "Q" does not express a fact that corresponds to an event, that fact has no causes.

There are also what Yablo calls enablers—"factors such that if you imagine them away, the cause . . . ceases to be enough for the effect": the presence of oxygen enables the striking of the match to be a cause of the lighting ("Advertisement," p. 98). By being an enabler, the presence of oxygen is a reason why ⟨the striking of the match was (causally) enough for the match to light⟩, without being a cause or a ground of this fact.[46]

For Yablo, ennoblers ennoble events into causes, and enablers enable events to cause their effects. But the concept of an ennobler and the concept of an enabler are more general than that. Facts can be ennoblers and enablers of *reasons*, not just of causes. If X is an ennobler of R with respect to the fact that Q, it ennobles R from "just another fact that obtains" in addition to the fact that Q, into a reason why Q. If X is an enabler of R with respect to the fact that Q, it enables R to be a sufficient reason why Q, when it otherwise would not be. Similarly, facts can be ennoblers and enablers of *grounds*.

Here's a crazy hypothesis:[47] *all* reasons why are either causes, grounds, *or ennoblers, or enablers*:

(T3) That R is a reason why Q iff the fact that R is a cause, or a ground, of the fact that Q, or, if the fact that Q is identical either to (i) the fact that X is a cause/ground of the fact that W, or to (ii) the fact that X is a reason why W, then the fact that R is an ennobler, or enabler, of the fact that X with respect to the fact that W.

I don't know if this is right—though it's better than (T2). And maybe further simplification is possible—you might think, contrary to what I've said, that ennobling and enabling are species of grounding. And I'm not sure how good (T3) will look when tested against claims about the reasons why certain mathematical truths are true. Still, it has a nice ring to it . . .[48]

---

[46] I said earlier in this chapter that the presence of oxygen was a reason why the striking was a reason; now I'm saying it's a reason why the striking was causally enough. I think both of these are true. I also complained in footnote 19 that the concept of an enabler was different from the concept of a reason why; the condition of my legs can enable me to be the team runner without being a reason why I'm the team runner. In that footnote I was using "enabler" in its ordinary sense; Yablo, I think, is using it in a semi-technical sense.

[47] Though I guess what makes it crazy is that it is the most conservative generalization from the examples we have seen.

[48] This appendix owes its existence to a conversation with Daniel Muñoz, Nathaniel Baron-Schmitt, and Quinn White.

Appendix C: Levels of Reasons and Grounding Necessitarianism

A second-level reason is a reason why something else is a reason. Next to second-level reasons I've often mentioned reasons why something is a cause. These are obviously closely connected; a reason why C is a cause of E is also, at least very often, a reason why C is a reason why E happened.

I didn't, however, mention reasons why something is a *ground*. But these things certainly exist. When G grounds H, we can ask why G grounds H, and an answer should provide some reasons why G grounds H.

My main claim about levels of reasons, HIGH$\nrightarrow$LOW, suggests some thoughts on the debate about "Grounding Necessitarianism." Grounding Necessitarianism is the thesis that, for any facts G1,..., Gn, and H, if G1,..., Gn (fully) ground H, then necessarily, if G1,..., Gn obtain, so does H. Some examples tell in favor of Grounding Necessitarianism: the fact that my backpack is red (fully) grounds the fact that my backpack is colored; and it is also true that, necessarily, if my backpack is red, then it is colored. But other examples are more problematic. What about, for example, dispositions? The fact that this piece of glass has categorical microstructure X (fill in the details) (fully) grounds the fact that it is fragile (doesn't it?); but it is false that, necessarily, if the glass has microstructure X, it is fragile. It could fail to be fragile, despite having microstructure X, if the laws of nature had been different.[49]

The usual response to examples that, like this one, appear to make trouble for Grounding Necessitarianism is to say that the facts listed do not really constitute a full ground. In this case, the response is that the fact that this piece of glass has microstructure X does not fully ground the fact that the glass is fragile. A full ground must include certain laws of nature, and those laws of nature, together with the microstructure, *will* necessitate the glass's fragility.

I'm wary of this response. I think we should all be wary of this response. The idea that whenever something is a ground there is a reason why it is a

---

[49] Many philosophers endorse Grounding Necessitarianism; for one example, see Gideon Rosen, "Metaphysical Dependence." Skiles, "Against Grounding Necessitarianism," argues against Grounding Necessitarianism, and provides a comprehensive list of proponents of the view in his footnote 3. Leuenberger, "Grounding and Necessity," argues that dispositions are a counterexample to Grounding Necessitarianism.

ground gives us something to say back. The laws of nature—certainly they are reasons why the microstructure grounds the fragility. Why should we demand any more of them than that? We shouldn't in the case of causation: certain laws of nature are reasons why Suzy's throwing the rock caused the window to break, but those laws are not also themselves causes of the breaking, nor are they (on my view) reasons why the window broke. Laws shouldn't have more privileges with respect to grounding than they do with respect to causation. We can let certain laws be reasons why the microstructure (fully) grounds the fragility, without requiring them to themselves be (partial) grounds of the fragility (and also without requiring them to be reasons why the glass is fragile).

My sense is that sympathy for Grounding Necessitarianism is partly driven by wonderment at how it could be false. If G *fully grounds* H, if H obtains *in virtue of the fact that* G obtains, how could it happen that G obtains without making H obtain? What is G not doing in a circumstance when H fails to obtain, that it *is* doing when G (fully) grounds H? I think the answer to this question is easy. G need not be doing anything different in a possible world in which it fails to ground H. What's different is not G but *the (actual) reasons why G grounds H*. The way for it to happen that G fails to ground H is not for G to "do something different" but for the reasons why G (actually) grounds H to fail to obtain (or to fail to be reasons).

5

Applications

5.1 Distinctively Mathematical Explanations

In a series of papers Marc Lange has championed the idea that there are non-causal explanations. In one of them, "What Makes a Scientific Explanation Distinctively Mathematical?," Lange tries to isolate a kind of explanation he calls "distinctively mathematical," and to argue that explanations of this kind are not causal explanations. I will spend a long time defending (T1) against the simplest of Lange's central examples; the simplicity of the example hides the complexity of the problems it raises for (T1), problems that can only be solved by discussing, not just first-level and second-level reasons why but third-level reasons as well. Here is the example:[1]

> STRAWBERRIES: "That Mother [let's call her Jane] has three children and twenty-three strawberries, and that twenty-three cannot be divided evenly by three, explains why [Jane] failed when she tried a moment ago to distribute her strawberries evenly among her children without cutting any [strawberries]." (p. 488)

Before thinking about what STRAWBERRIES says are the reasons why Jane failed, let's figure out what (T1) says are the reasons. First, what are the causes of her failure? The fact that Jane has (exactly) three children is certainly a cause of her failure to distribute the strawberries evenly, as is the fact that Jane has (exactly) twenty-three strawberries. (Each passes a counterfactual test for causation: if she had twenty-three children instead,

---

[1] A third-level reason why, of course, is a reason why ⟨A is a reason why ⟨B is a reason why C⟩⟩, for some A, B, and C. Lange takes the example from David Braine's paper "Varieties of Necessity." Lange does not discuss (T1) specifically (he couldn't have), though I believe he would take his example to refute (T1); anyway, I will reinterpret his arguments that the example is not a causal explanation as arguments that it is a counterexample to (T1).

she wouldn't have failed; if she had twenty-four strawberries instead, she wouldn't have failed.[2]) So, according to (T1), these facts are reasons why Jane failed. The (purely mathematical) fact that three does not divide twenty-three, on the other hand, does not cause anything, and so is not a cause of Jane's failure. Nor does this mathematical fact ground her failure. According to (T1), then, this mathematical fact is not a reason why Jane failed.

STRAWBERRIES can be read as entailing the contrary, as entailing that the fact that three does not divide twenty-three *is* a reason why Jane failed. I'm not sure whether it really does entail this; all that STRAWBERRIES asserts explicitly is that this mathematical fact belongs to a collection of facts that (collectively) "explain why" Jane failed. Whether this entails that the mathematical fact is a *reason why* she failed depends on whether a collection of facts that "explain why Q" must be such that each of the facts in the collection constitutes a reason why Q. I will interpret STRAWBERRIES as making this claim, since only then is the example a challenge to (T1).

Is STRAWBERRIES, read this way, true? Is the fact that three does not divide twenty-three a reason why Jane failed to distribute the strawberries equally? Why would anyone think that it was? I take it that the only temptation to accept that it is a reason comes from the naturalness of this dialogue:

A: Why did Jane fail to distribute the strawberries equally?
B: She had three children, and twenty-three strawberries, and three does not divide twenty-three.

I want to say the same thing about this dialogue that I said about the SECOND ROCK DIALOGUE from the last chapter:

A: Why did the rock hit the ground at a speed of 4.4 m/s?
B: It landed at a speed of 4.4 m/s because I dropped it from 1 meter, and Newton's theory of gravitation entails that, for short falls, the impact speed s is related to the distance fallen d by the equation $s = \sqrt{2dg}$, where g is the gravitational acceleration near the surface of the earth. And of course $\sqrt{2 \cdot 1 \cdot 9.8} \approx 4.4$.

---

[2] Lange accepts that these facts are causes; see p. 495.

On my view, the law that $s = \sqrt{2dg}$ appears in B's response, but is not part of B's answer to A's question. It is instead an answer to an unasked vertical follow-up question. It is offered as a second-level reason, a reason why ⟨the 1 meter drop height is a reason why the rock landed at 4.4 m/s⟩, not as a reason why the rock landed at 4.4 m/s. Similarly, the fact that three does not divide twenty-three is not part of B's answer, but is offered as a second-level reason why the facts about the number of children and of strawberries are reasons why she failed. Including only the reasons why Jane failed is apt to leave one's audience unsatisfied. The best way to avoid this is to include, in one's response, the reason why the reasons one is offering are reasons.

Lange says some other things about "distinctively mathematical explanations" that make them a challenge for me, even if I am right that the fact that three does not divide twenty-three is not a reason why Jane failed. (T1) does not just say that every cause of E is a reason why E happened. It says that for each cause of E, the *reason why it is a reason* why E happened is that it is a cause of E. Applied to this case, (T1) says that the reason why the fact that Jane had three children is a reason why she failed is that it is a cause of her failure. Lange, I think, would deny this claim.[3] "Even if [these answers] happen to appeal to causes," Lange writes, "they do not appeal to them as causes" (p. 496). I think it is best to abstract away from what any particular answer does or does not do, and just focus on what reasons there are. The claim about reasons why that is closest to the claim Lange makes in the quotation is this one: even if some causes are reasons why E happened, the reasons why those causes are reasons are not that they are causes.

Is this a plausible thing to say about STRAWBERRIES? Why deny that ⟨the fact that Jane's having three children caused her failure⟩ is the reason why ⟨her having three children is a reason why she failed⟩? Lange does not take up precisely this question, but looking at the positive suggestion he makes after he denies that STRAWBERRIES "appeals to causes as causes" points toward an answer:

> Even if they [examples like STRAWBERRIES] happen to appeal to causes, they do not appeal to them as causes—they do not exploit

---

[3] I discussed this strategy of his for arguing against (T1) in chapter 3, when I tried to understand Lange's talk of "sources of explanatory power."

their causal powers. In particular, I now suggest, any connection they may invoke between a cause and the explanandum [the event the why-question concerns] holds not by virtue of an ordinary contingent law of nature, but typically by mathematical necessity.

(p. 496–7)

Here is one way to turn this suggestion into an argument that the reason why ⟨Jane's having three children is a reason why she failed⟩ is not that her having three children caused her failure (hold on tight; I do not know how to express it without difficult-to-parse iterations of "reason why"):

Let's focus on just one of the reasons why ⟨Jane's having three children is a reason why she failed⟩, namely the fact that three does not divide twenty-three. What is interesting and important about this reason is that it is a mathematical fact, instead of a "causal law." Now I hold that, in general, if a mathematical fact is the only "nomological" reason why ⟨C is a reason why E happened⟩, then even if C is a cause of E, it is false that ⟨the fact that C is a cause of E⟩ is a reason why ⟨C is a reason why E happened⟩. The condition is satisfied in this case; the fact that three does not divide twenty-three is the only nomological reason why ⟨Jane's having three children is a reason why she failed⟩. That is why, even though Jane's having three children is both a reason why she failed and cause of her failure, it is not a reason because it is a cause.

What do I mean, "nomological reason"? I just mean a reason that is a law, or at least law-like, if mathematical truths do not count as laws.

The fact that three does not divide twenty-three is not the only reason why ⟨Jane's having three children is a reason why she failed⟩. There is also the fact that she had twenty-three strawberries. But this other fact is not a nomological reason, for it is nothing like a law. It is instead a fact that corresponds to the occurrence of an event. Its being a reason why ⟨Jane's having three children is a reason why she failed⟩ just reflects the fact that, when an event has many causes, each cause is typically a reason why each other cause is a reason why the event happened.

If the nomological reason why ⟨Jane's having three children is a reason why she failed⟩ *had* been some "causal law," like say Newton's law of universal gravitation, then it would have been different. When the relevant reason why ⟨C is a reason why E happened⟩ is a causal law, then the fact that C is a cause of E *is* a reason why ⟨C is a reason why E happened⟩. But that's not what's going on in this case.

The key premise in this speech is

- If every nomological reason why ⟨C is a reason why E happened⟩ is a mathematical fact, then even if C is a cause of E, it is false that the fact that C is a cause of E is a reason why ⟨C is a reason why E happened⟩.

But other things Lange says in his paper suggest that he would want a slightly different premise, one that allows other kinds of nomological reasons, besides mathematical facts, to generate counterexamples to (T1). For he thinks that you also get a "distinctively mathematical explanation" when the connection invoked between the cause and effect is, not a mathematical truth, but a contingent law of nature that is "more necessary" than the causal laws, in the sense that that law still would have held had the causal laws been different. These "more necessary" laws—we can call them "higher-order" laws of nature—can be thought of as placing constraints on what the causal laws are, as providing the framework that the causal laws inhabit. Here is Lange:

> The natural laws in a distinctively mathematical explanation in science, I suggest, must transcend the laws describing the particular kinds of causes there are. A distinctively mathematical explanation in science works not by describing the world's actual causal structure, but rather by showing how the explanandum [the event the why-question concerns] arises from the framework that any possible causal structure must inhabit, where the 'possible' causal structures extend well beyond those that are logically consistent with all of the actual natural laws there happen to be.   (p. 505)

Newton's second law is a plausible example of a higher-order law. Plausibly, it still would have been a law even if the force laws had been different—even if, say, gravity had been an inverse cube law instead of an inverse square law. Newton's second law seems to be part of a "framework that any possible causal structure must inhabit," in the sense that it requires (basic) causes to act by generating forces on things, which then cause those things to accelerate. An answer to a why-question that cites causes, but invokes only Newton's second law to connect those causes to their effect, counts as a distinctively mathematical explanation.

These claims about higher-order laws suggest that our key premise should not be the one I wrote down above, but

- If every nomological reason why ⟨C is a reason why E happened⟩ is either a mathematical fact *or a higher-order law of nature*, then even if C is a cause of E, it is false that the fact that C is a cause of E is a reason why ⟨C is a reason why E happened⟩.

But actually I'm going to say that this isn't quite the premise we want either. Mathematical facts and higher-order laws have in common that they are "more necessary" than the ordinary causal laws. But they're not the only necessities that are more necessary. Metaphysical necessities are as well. I see no reason to say that an answer to a why-question can get to be "distinctively mathematical" by citing a mathematical truth or a higher-order law, but could not get to be distinctively mathematical by citing a metaphysical necessity. So here is the final version of the key premise:

KEY PREMISE: If every nomological reason why ⟨C is a reason why E happened⟩ is either a mathematical fact, or a metaphysically necessary truth, or a higher-order law of nature, then even if C is a cause of E, it is false that the fact that C is a cause of E is a reason why ⟨C is a reason why E happened⟩.

Is the argument that STRAWBERRIES is a counterexample to (T1) sound? Is KEY PREMISE true? I will start working to undermine the premise by changing the subject, to Zeno causality.[4] Imagine a line with a coordinate system on it, the point labeled 0 "in the middle," points labeled with negative numbers on the left, points labeled with positive numbers on the right. A ball rolls along the line from left to right. The ball must move continuously through space; it cannot jump discontinuously from one place to another. It has not yet reached the positive numbers. There are infinitely many walls (each infinitely thin), labeled with whole numbers (so the first wall is labeled with 0), each in its "down" position, arrayed along the line between the point labeled 0 and the point labeled 1. The

---

[4] See, for example, John Hawthorne's paper "Before-Effect and Zeno Causality."

walls are the only things around other than the ball. When a wall is "down" the ball will roll right over the top of it. But when it is up it has the power—never mind how, or by what causal laws—to cause any ball that reaches it to stop. Wall 0 is located at point 1, wall 1 is located at point 1/2, wall 2 is located at point 1/4, wall n is located at point $1/2^n$; I will say that the walls are arranged in a "ω-sequence" ("reverse-omega sequence"). Each wall is programmed to very quickly move to its up position if, but only if, the leading edge of a ball makes it halfway to it from the location of next wall. So wall n goes up iff a ball's leading edge makes it half the distance from the position of wall $n + 1$—a distance of $\frac{1/2^n - 1/2^{n+1}}{2}$ units—and it goes up fast enough that it has made it all the way up before the ball gets to it.

What will happen? The ball will stop before it gets to any wall. For the description of the situation is to be understood to entail the following:

(Z1)   For each n, if wall n goes up, the ball does not get past wall n.

(Z2)   For each n, if wall n goes up, then wall $n + 1$ went up, and the ball got past wall $n + 1$.

(Z1) and (Z2) entail, for each n, that wall n does not go up. Since every wall is wall n for some n, we have that no wall goes up. Now

(Z3)   If the ball ever occupies a point with a positive coordinate, then some wall goes up.

So we may conclude that the ball never occupies a point with a positive coordinate. It stops at some point with a non-positive label.[5]

What (thing) caused the ball to stop? Something did: the laws governing the situation (let me stipulate) do not permit the ball to stop "spontaneously," for no reason at all. But no individual wall caused the ball to stop. For no wall exercised its power to cause balls to stop—they can only exercise this power in the up position. What else is there? The example is to be understood so that there are no other obvious candidates: there's nothing distinct from the walls around to stop the ball; the ball did not "stop itself"—it's not disposed to stop when it gets to *that place* (the place where it stopped), nor does it have any similar disposition, like a disposition to stop every ten seconds. Hawthorne does not discuss exactly this example, but it is clear that he would say that the *fusion* of the walls

---

[5] Maybe it stops *and turns around*; what I have said underdetermines whether it does this, or stops and remains stopped.

(the smallest material thing that has each of the walls as a part) caused the ball to stop (see p. 630).

The thesis that the fusion of the walls caused the ball to stop raises a lot of interesting questions. Hawthorne's interest is in answering this question: *how* did the fusion cause the ball to stop? He would claim that the fusion didn't do anything, beyond being arranged in a ω-sequence, to cause the ball to stop.[6] Certainly this fact passes a simple (but of course defeasible) counterfactual test for causation: if the walls hadn't been arranged in a ω-sequence, the ball wouldn't have stopped.[7] But for us Hawthorne's how-question is less interesting than a why-question: *why* was the walls' being arranged in a ω-sequence a cause of the stopping? What are the reasons why the fact that the walls were arranged in a ω-sequence was a cause of the stopping? I think that the only nomological reason is this:

> BESTOWAL: It is metaphysically necessary that, if some individual walls have, by virtue of some causal law or other, the power, when up, to cause any ball that reaches them to stop—so that (Z1) is true; and if those walls are arranged in a ω-sequence, are all in the down position, and are programmed, it doesn't matter how, in a way that makes (Z2) true; then the *fusion* of those walls will cause any ball that threatens to reach it to stop.

BESTOWAL is a metaphysical necessity that articulates a framework that any causal structure must inhabit. A set of causal laws that grants some walls individually a certain power must, if certain other conditions are met, bestow upon the fusion of those walls a related causal power. As I said, I think that BESTOWAL is the only nomological reason why the fact

---

[6] The moral he draws from examples like this one is that the following principle is false: "if x is the fusion of y's and y's are individually capable only of producing effect e by undergoing change, then x cannot . . . produce effect [e] without undergoing change" (p. 630).

[7] Well, there are lots of ways to arrange the walls other than in a ω-sequence, and there are some ways such that, had the walls been arranged in one of them, the ball still would have stopped (for example, if not all but an infinite subset of the walls had been arranged in a ω-sequence, or if some of the walls had started in their up position). But there are also some alternative ways of arranging the walls such that, had they been arranged that way, the ball would not have stopped—for example, had wall $n + 1$ been located to the right, instead of the left, of wall n. This more sophisticated counterfactual is the one I am using to test for causation. If you don't think the truth of this counterfactual follows from the example as so far described, let it be part of the description.

that the balls were arranged in a ω-sequence was a cause of the stopping; if it is, then we have here a case where a metaphysical necessity is a reason why some fact C (the fact that the walls are arranged in a ω-sequence) is a cause of a fact E (the fact that the ball stopped).

Please do not think that I have said that BESTOWAL is the only reason *of any kind* why the fact that the balls were arranged in a ω-sequence was a cause of the stopping. I have said only that it is the sole nomological reason why. There are certainly other reasons: certainly the fact that each wall had the power, when up, to stop any ball that reached it, is also a reason. But this fact, concerning as it does the powers of a definite collection of individuals, corresponds to an event; it is not a law.[8]

Let me spell out slowly how Zeno causation, and the claims I have made about it, undermine KEY PREMISE. In the case of Zeno causation the following are, I believe, true:

(1) The fact that the walls were arranged in a ω-sequence was a cause of the stopping.

(2) The fact that the walls were arranged in a ω-sequence is a reason why the ball stopped.

(3) BESTOWAL is a reason why (1) is true, that is, is a reason why ⟨the fact that the walls were arranged in a ω-sequence was a cause of the stopping⟩.

(4) BESTOWAL is (also) a reason why (2) is true, that is, is a reason why ⟨the fact that the walls were arranged in a ω-sequence is a reason why the ball stopped⟩.

Now I think that the truth of (3) undermines KEY PREMISE as applied to this case. Why should we conclude, from the fact—in (4)—that a certain metaphysical necessity is the only nomological reason why the cause is a reason, that it is false that the cause is a reason because it is a cause, as KEY PREMISE instructs us to do? (More slowly: why should we conclude, from the fact that a certain metaphysical necessity is the only nomological reason why ⟨the arrangement of the walls is a reason why the ball stopped⟩, that it is false that a reason why ⟨the arrangement of the walls is a reason why the ball stopped⟩ is that the arrangement was a cause of the

---

[8] This comment might inspire one to look for other nomological reasons; I will say something more about why there aren't any below, when I respond to an objection.

stopping?) All I can come up with is this: (i) that metaphysical necessity is certainly not a reason why the cause is a *cause*; so (ii) if that metaphysical necessity is the only reason why the cause is a *reason*, the fact that the cause is a cause is "irrelevant," it has nothing to do with why the cause is a reason.

That, as I said, is the only line of thought I see for thinking that KEY PREMISE is right about this case. But the line of thought rests on a false premise. Claim (i) in this case is false. That metaphysical necessity *is* a reason why the cause is a cause. KEY PREMISE is, I conclude, wrong about this case.

Objection: "Your argument against KEY PREMISE rests on the claim that BESTOWAL is the *only* nomological reason why the fact that the walls are arranged in a ω-sequence was a cause of the stopping. But it's not. The ordinary, 'first-order' causal laws are also nomological reasons why this is a cause." No, they are not. I haven't said what the causal laws of the Zeno world are; but whatever they are, they attribute causal powers to individual walls, not to walls collectively arranged in ω-sequences. More importantly, facts about what the ordinary causal laws are are "too specific" to be reasons why the fact that the walls are arranged in a ω-sequence is a cause of the stopping. For those causal laws will specify, as it were, the exact causal mechanism—be it a familiar one, like electromagnetism, or some completely alien one—by which walls have the power to stop approaching balls when in the up position. But the fact that the walls were arranged in a ω-sequence would still have caused the stopping, had the causal laws been different, had the mechanism by which the walls had this power been different. What matters is not the laws that give the individual walls their causal powers but the constraint on the distribution of causal powers that BESTOWAL articulates.

What I have said about Zeno causation, I want to say about STRAW-BERRIES. Let C be the fact that Jane has three children. I agree that the fact that three does not divide twenty-three—a mathematical truth—is a reason why, and is the only nomological reason why, C is a reason why she failed to distribute her strawberries evenly. But KEY PREMISE is not in general true, and is not true of this case. It does not follow that the fact that ⟨C is a cause of her failure⟩ is *not* a reason why ⟨C is a reason why she failed⟩. In fact, I think that the mathematical fact that three does not divide twenty-three is not just a reason why C is a reason but also a reason why C is a cause; and in light of this we can conclude that one reason why C is a reason is that it is a cause.

Lange had, I think, a tremendously important insight into examples like STRAWBERRIES. Even though the answer to the question why Jane failed mentions causes of her failure, the relationship between those causes and her failure is different from the "usual" relationship between causes and effects. Recall his observation that the connection that answers like STRAWBERRIES "invoke between a cause and the explanandum [the event we have asked about] holds not by virtue of an ordinary contingent law of nature, but typically by mathematical necessity" (pp. 496–7). I think this is absolutely right. I don't think it undermines (T1).

The correct moral to draw is not about "explanation," or answers to why-questions, but about causation. The moral is *not* that, sometimes, even when C is a cause of E, its being a cause is not a reason why ⟨it is a reason why E happened⟩. The moral instead is that, sometimes, the reasons why C is a cause of E do not include, or do not only include, ordinary causal laws, or contingent laws of nature. Sometimes the reasons why C is a cause of E are "more necessary" than such laws—they are either higher-order laws of nature, or mathematical truths, or metaphysical necessities more generally.

5.2 Third-Level Reasons

There is an objection I must discuss, that I have been trying to avoid. I have been trying to avoid it because responding to it requires discussing third-level reasons—answers to questions like: why is A a reason why ⟨B is a reason why C⟩? Thinking about second-level reasons is hard enough; thinking about third-level reasons requires good daylight and a lot of caffeine.

There are benefits to discussing the objection (in addition to showing it to fail). Addressing it will let me sharpen up the argument I have been giving against KEY PREMISE.

The objection is that everything I have said in response to the strawberries example presupposes a claim that is itself inconsistent with (T1). My attempts to show that the strawberries example is consistent with (T1) has involved implicitly abandoning (T1) from the beginning. The cure is worse than the disease.

What is that presupposition? This:

(5) The fact that three does not divide twenty-three is a reason why ⟨the fact that Jane had three children is a reason why she failed⟩.

I asserted this right away, as an alternative to the (in my view) false claim that the mathematical fact is a (first-level) reason why Jane failed.

But look again at the part of theory (T1) that is about causes (so I am omitting all mention of grounding):

Necessarily, if it is a fact that Q and it is a fact that R, then: if the fact that R is a cause of the fact that Q, then one reason why Q is that R, and the reason why ⟨one reason why Q is that R⟩ is that the fact that R is a cause of the fact that Q.

The crucial bit in this is where it says that *the* reason why ⟨one reason why Q is that that R⟩ is that the fact that R is a cause of the fact that Q. *The* reason. (T1) therefore entails that, for any cause, there is only *one* reason why that cause is a reason why its effects happen, namely the fact that it is a cause. But I have many times so far admitted the existence of *other* reasons why a cause is a reason. Claim (5) contains an example. So the truth of (5) is inconsistent with (T1) (on the assumption that Jane's having three children is a cause of her failure).

And (5) is not an isolated example. A similar claim is true about the Zeno causation example. I said that a certain metaphysical necessity is a reason why ⟨the arrangement of the walls is a reason why the ball stopped⟩; so I admit that the fact that the arrangement is a cause is not the only reason why the arrangement is a reason. And throughout chapter 4 I said that many facts (that are not themselves causes) that have been taken to be reasons why some events occurred are really just reasons why the causes of those events are reasons.[9]

---

[9] You might question why I said this. Why didn't I just say that those facts are reasons why those causes are *causes*? With respect to STRAWBERRIES, the question is: why didn't I just say that the fact T that three does not divide twenty-three is (only) a reason why Jane's having three children is a *cause* of her failure, and deny that T is a reason why her having three children is a reason why she failed? Isn't it enough for T to be a reason why the cause is a cause? Why think it is also a reason why the cause is a reason?

If NON-TRANSITIVITY were false, if "reason why" were transitive, this question wouldn't arise; given that T is a reason why C is a cause, and C's being a cause is a reason why C is a reason, both of which I accept, it would *follow* that T is a reason why C is a reason. But NON-TRANSITIVITY is true.

It is exactly because of truths like (5) that I held back and did not officially endorse (T1). It is exactly because of truths like (5) that I do not want to say that (T1) is the absolutely correct way to capture the idea that "explaining events is about describing causes." Fidelity to this idea does not, I think, require one to go so far as to say that the *only* reason why a cause is a reason is that it is a cause. All it requires one to say is that *there are no reasons why a cause is a reason that are, as it were, "completely independent" of the fact that it is a cause.* I have been implicitly relying on something like this idea in my discussion of the strawberries and Zeno examples, but it is time to make it precise and fully explicit. Here, then, is the final version of my theory:

(T1f)   Necessarily, if it is a fact that Q and it is a fact that R, then: (i) if the fact that R is a cause of the fact that Q, then one reason why Q is that R; and (ii) *one* reason why ⟨one reason why Q is that R⟩ is that the fact that R is a cause of the fact that Q; moreover, (iii) for *every other* reason X why ⟨one reason why Q is that R⟩, the only reason why ⟨X is a reason why ⟨one reason why Q is that R⟩⟩ is that ⟨X is a reason why the fact that R is a cause of the fact that Q⟩. (And similarly for grounding; and there are no reasons why Q other than causes and grounds of the fact that Q that satisfy these conditions.)

With (T1) amended this way, what I have said about STRAWBERRIES, and about Zeno causation, is consistent with it. Yes, certain mathematical, or metaphysical, necessities are reasons why the causes in those scenarios are reasons. But they are only reasons why the causes are reasons because they are reasons why the causes are causes.

Look again at claims (3) and (4) that I made about Zeno causation:

(3)   BESTOWAL is a reason why ⟨the fact that the walls were arranged in a ω-sequence was a cause of the stopping⟩.

(4)   BESTOWAL is a reason why ⟨the fact that the walls were arranged in a ω-sequence is a reason why the ball stopped⟩.

Honestly, it occurred to me only just before sending off this book to be published that denying that T is a reason why the cause is a reason was an option worth considering. I guess in these cases it struck me as right that certain facts were *both* reasons why the cause was a cause and reasons why the cause was a reason, despite the fact that "reason why" is not in general transitive.

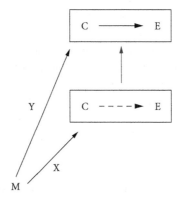

Figure 5.1. diagram for (6)

Surely there is a connection between (3) and (4). Isn't (3) the answer to the question why (4) is true? Here is the question: why is it that BESTOWAL is a reason why ⟨the arrangement of the walls is a reason why the ball stopped⟩? It couldn't be irrelevant that BESTOWAL is (also) a reason why ⟨the arrangement of the walls is a *cause* of the stopping⟩! Indeed, I hold that (4) is true because (3) is true; (3) and (4) are not true "independently."

Now I can strengthen my argument against KEY PREMISE, for I think that the following proposition is true:

(6)   If M is a reason why ⟨C is a reason why E happened⟩ only because M is a reason why ⟨C is a cause of E⟩, then ⟨the fact that C is a cause of E⟩ is a reason why ⟨C is a reason why E⟩.

It is easier to see the appeal of (6) if we picture what is going on. Figure 5.1 has the situation: M is the mathematical or metaphysical law, or the higher-level law of nature that transcends the causal laws; C is the cause, E is the effect. Solid arrows indicate reasons why, dashed arrows indicate causation. In the top box we have the fact that C is a reason why E happened. In the box directly below we have the fact that C is a cause of E. Arrow X indicates that M is a reason why C is a cause of E. Arrow Y indicates that M is a reason why ⟨C is a reason why E happened⟩.

I have said that, in the Zeno example, arrow Y is there only because arrow X is there. (That is: claim (3) gives the reason why (4) is true.) But for that to be the case, shouldn't there be an arrow from the lower box to the top box? How can the arrow from M to the top box be there only because the arrow from M to the lower box is there, if there is no arrow from the lower box to the top box? If M can "get to the top box" only because it can "get to the lower box," mustn't one be able to get from the lower box to the top box? (If one can get to New York only because one can get to New Haven, mustn't one be able to get from New Haven to New York?) I think that the answer to this last question is yes. But the existence of this additional arrow means that the fact that C is a cause of E is a reason why ⟨C is a reason why E happened⟩. And that is what was to be shown. So (6) is true. But then cases that satisfy its antecedent—the Zeno example and STRAWBERRIES are such cases—are counterexamples to KEY PREMISE.

That was exhausting. While we rest and recover I'll go through another, less exotic, counterexample to KEY PREMISE.

Robin, a ball of unit mass, is at rest (Robin is initially at rest), while Martin, a second ball of unit mass, moves at unit speed toward Robin along a line (a line that passes through the centers of both balls), in some direction D, on a flat frictionless surface. What happens? Martin collides with Robin, comes to rest, and Robin moves off at unit speed in direction D. Focus on a moment after the impact. Here is the question: why is Robin moving with unit speed in direction D? This much seems uncontroversial: one reason why is that

(B1)   Robin was hit by Martin, an equally massive ball moving with unit speed in direction D.

Another reason why is that

(B2)   Robin was at rest before the impact.

These reasons are causes; I hold, further, that they are reasons because they are causes.

I think my judgment here—a judgment that is, of course, in accord with my theory—is not radical or controversial.[10] Isn't this just an ordinary example of "causal explanation"? *I* think so, but if we look more closely at the details of the example, this judgment looks to be incompatible with KEY PREMISE. For what are the nomological reasons why (B1) and (B2)

---

[10] Here I'm just asserting that the causes are reasons because they are causes, and expecting you to agree. If an argument is needed, one can be constructed, using thesis (6) as a premise, as well as the claims I am about to make about the nomological reasons why (B2) and (B2) are reasons.

are reasons why Robin moved off at unit speed in direction D? They are the following:

(H1) Energy is conserved.

(H2) Momentum is conserved.

(H3) No part of space is ever simultaneously occupied by parts of both balls.

(H4) The balls move continuously through space.

(H5) Balls continue moving at the same speed in the same direction unless doing so conflicts with one of the laws (H1) through (H4) (this is a modified version of the law of inertia).

(H6) If, in a closed system composed of two balls, there is a time at which the motion of both balls is confined to a single line (each ball is either at rest, or moving along that line), then their motion is always confined to that line (so long as they remain a closed system).

Given Robin and Martin's initial states, there are only two states of motion of the balls that are consistent with (H1), (H2), and (H6): the state S1 in which Robin is at rest and Martin moves at unit speed in direction D, and the state S2 in which *Martin* is at rest and Robin moves at unit speed in direction D.[11] By hypothesis the balls start in S1. By (H5) they stay in S1 until Martin gets to a part of space occupied by Robin. Laws (H3) and (H4) entail that the balls cannot remain in state S1 past that time, so they must go into (and by (H5), remain in) state S2, which is what was to be shown.

Laws (H1) through (H6) are not mathematical truths but we may take them to be higher-order laws of nature. Lange himself is certainly sympathetic to the idea that conservation laws, like (H1) and (H2), and symmetry principles, like (H6), are higher-order laws.[12] What about (H3)? (H3) is certainly true of actual medium-sized material bodies but I doubt that (H3) is a higher-order law. Nevertheless, I think we may *take it to be* one for purposes of the example. Let us just suppose for purposes of

---

[11] If V_r and V_m are Robin's and Martin's velocities, with positive numbers representing motion in direction D, and negative numbers motion in the opposite direction, then the law of energy conservation requires $V_r^2 + V_m^2 = 1$, while the law of momentum conservation requires $V_r + V_m = 1$. The only two solutions are $V_f = 1, V_m = 0$ and $V_f = 0$, $V_m = 1$.

[12] See his paper "Laws and Meta-laws of Nature: Conservation Laws and Symmetries."

the example that it is in the nature of material things to exclude other material things from the regions of space they occupy. (This was certainly thought to be part of the nature of material things at one point in the history of natural philosophy.[13]) Moving on to (H4), the requirement of continuous motion, it also seems, like the conservation laws, to be not just a law governing material bodies but a law constraining what the first-order causal laws are. (Or, if (H4) is actually false, it may be taken to be such a law.) Similarly, we make take (H5), the modified law of inertia, to be a higher-order law. (It is certainly eligible to be a higher-order law—like the real law of inertia, and like Newton's second law, it is the kind of thing that might well place constraints on what the first-order causal laws are.)

There are a lot of ifs and suppositions behind the claim that (H1) through (H6) may be taken to be higher-order laws of nature. But none of them are implausible. I am not going way out on a limb.

KEY PREMISE, then, entails that (B1) (Martin's colliding with Robin), while a reason why Robin is moving, and a cause of Robin's motion, is not a reason because it is a cause; and similarly for (B2). But that's not right. KEY PREMISE is false.

5.3 Presupposition, Necessity, Coincidence

So far, everything I have said in this chapter has been aimed at responding to a single, seemingly very simple, example, namely STRAWBERRIES. What I would say about many other examples can be gleaned from my discussion of it. But I am not going to rest the defense of my theory on the consideration of a single case, not least because other examples raise complications that have not yet come up.

So here is another of Lange's central examples of a distinctively mathematical explanation:[14]

BRIDGES: "Why did a given person [say, Jones] on a given occasion not succeed in crossing all of the bridges of Königsberg exactly once (while remaining always on land or on a bridge rather than in a boat,

---

[13] Henry More (d. 1687) is one example of a philosopher who accepted this claim. For discussion of the historical context, see Des Chene, *Physiologia*, pp. 377–82.

[14] Pincock also discusses this example in detail in his paper "A Role for Mathematics."

for instance, and while crossing any bridge completely once having begun to cross it)?... The distinctively mathematical explanation is that in the bridge arrangement, considered as a network, it is not the case that either every vertex or every vertex but two is touched by an even number of edges. Any successful bridge-crosser would have to enter a given vertex exactly as many times as she leaves it unless that vertex is the start or the end of her trip. So among the vertices, either none (if the trip starts and ends at the same vertex) or two could touch an odd number of edges" (pp. 488–9).

Let's read Lange as saying that one *reason why* Jones failed to cross all the bridges exactly once is that "in the bridge arrangement, considered as a network, it is not the case that either every vertex or every vertex but two is touched by an even number of edges." A graph, in the mathematician's sense, is a set of vertices (or nodes) and edges, where each edge touches two vertices, together with a specification of which edges touch which vertices. An Eulerian graph is one in which every vertex is touched by an even number of edges. A semi-Eulerian graph is one in which every vertex but two is touched by an even number of edges. The fact Lange cites to answer the why-question, then, is the fact that the bridge-and-landmass arrangement in Königsberg lacks property P: the property of being representable by either an Eulerian or semi-Eulerian graph. Lacking P is a perfectly respectable physical property of the bridge set-up, even if it is a relatively non-specific property (it can be instantiated in a great number of physically different ways) that has been picked out using graph-theoretic language.[15] The fact that the bridge set-up lacks P is clearly a cause of Jones's failure. He would not have failed if the bridge set-up had had P instead. I think that this cause is a reason why Jones failed, and that it is a reason because it is a cause; I think that the example is consistent with (T1f) (I will not explicitly go through all the reasons why the bridge set-up's lacking P is a reason why Jones failed).

Again, I interpret Lange as claiming that the fact that the bridge set-up lacked P, while both a cause of Jones's failure and a reason why he failed, is

---

[15] One of Pincock's original points about this example was just that what made the graph-theoretic language useful was that it allowed us to find the right level of "abstraction" for answering the question why Jones failed. (However, the graph-theoretic language does not seem to be essential. We could just as easily say that it is false that either every landmass, or every landmass but two, is touched by an even number of bridges.)

not a reason because it is a cause. We have been over this ground before. But a few new things also come up.

Lange says at one point that "the fixity of the arrangement of bridges and islands . . . is presupposed by the why question that the explanation answers" (p. 497). We could make this presupposition explicit in the question:

(7)   Why, *given that* the bridge set-up lacked P (and continued to lack P throughout Jones's attempt), did Jones fail?

An answer to this question cannot cite the fact—call it fact F—that the bridge set-up lacked P, since in general one cannot answer the question why Q by providing a reason why Q when that reason was presupposed by the question in the first place. "I know that the window was fragile, and that its fragility is one reason why it broke, but please tell me, why did the window break?" You cannot say, "Because it was fragile."

Focusing on the question-with-presupposition (7) changes things. Fact F is the only relevant cause of Jones's failure. So if the answer in BRIDGES is an answer to (7), that answer cannot "work" by providing F as a reason why Jones failed. It must provide some other reason; surely that reason will have to be a fact that was not a cause. What fact is this? It must be the fact that "[a]ny successful bridge-crosser would have to enter a given vertex exactly as many times as she leaves it unless that vertex is the start or the end of her trip. So among the vertices, either none (if the trip starts and ends at the same vertex) or two could touch an odd number of edges." Let me call this fact "fact G" and restate it as: the fact that, necessarily, if the bridge set-up lacks P, then no one crosses each bridge exactly once. So the claim I need to worry about is the claim that G is offered in BRIDGES as a reason why Jones failed.

Before we focused on the question-with-presupposition (7), I would have said that G is not a reason why Jones failed, it is just a reason why the fact that the bridge set-up lacked P is a cause of the fact that, and a reason why, Jones failed. I continue to say that now. How is this consistent with the fact that asserting that G obtains is a good thing to do in response to (7)? After all, (7) is a request for reasons why Jones failed.

(7) is a weird question. It does not wear its weirdness on its face, but I think that the situation in which it is asked is analogous to situations like this. You and I watch Suzy throw a rock at a window. We watch the rock break the window. We both see that there is no one else around, and see that the other sees this. You turn to me and ask, "Why did the window

break?" Not sure what is going on, I reply, "…because Suzy threw a rock at it …" You say, "yes yes yes, I know that; what I want to know is, *given that* Suzy threw a rock at the window, why did it break?"

It's hard to know what to say. The question is weird. You have explicitly verified for me that you know, of the only relevant cause of the breaking, that it was both a cause of the breaking and a reason why the window broke.[16] You have asked why the window broke. But there's nothing more to tell you. There are no other reasons why the window broke.

I think that it is natural, when someone asks a weird why-question like this, one where the asker manifestly already knows the answer, to reinterpret the question. For there is a why-question nearby that you might not know the answer to. You might not know the answer to the question, why is the fact that Suzy threw a rock a cause of the fact that (or: a reason why) the window broke? To interpret the question this way is to hear "Given X, why Q?" as a way of asking why X is a reason why Q.

With respect to (7), the reinterpretation has it asking why the fact that the bridge set-up lacked P is a reason why Jones failed. And to *this* question fact G is certainly an answer, even on my view.

My diagnosis, then, is this: (7) can be heard to ask two different questions, one requesting reasons why Jones failed other than the presupposed reason, one requesting reasons why the presupposed reason is a reason. Lange is right that G can be correctly offered as a reason in response to one of those questions—but only to one of them. The question to which it can correctly be offered as a reason is the request for a second-level reason.

Early in his paper Lange puts the answers to the questions why Jane failed and why Jones failed like this: "The Königsberg bridges as so arranged were never crossed because they *cannot* be crossed. [Jane]'s strawberries were not distributed evenly among her children because they *cannot* be" (p. 491; original emphasis). These answers look like counterexamples to (T1f); the impossibility of crossing the bridges cannot cause anything, so is not a cause of anyone's failure to cross the bridge. After thinking through what to say about the answer to (7) I now know what to say about these answers.

The "cannots" signal impossibility, and on the surface the sentence "Jane's strawberries were not distributed evenly among her children because they cannot be" appears to say that distributing the strawberries

---

[16] If you think there are other relevant causes (that the window is not super-strong, that nothing interfered with the rock mid-flight . . .), add them in to what we both know.

evenly was unconditionally impossible. But of course it was not, nor would anyone say it was. Distributing the strawberries evenly is only "conditionally" impossible, impossible given that, and holding fixed the fact that, she had three children and twenty-three strawberries. Since the conditions are not on the surface, they must be hidden somewhere, and where they are hidden is in the question being answered. "Because distributing them evenly was impossible" does not answer the question why Jane failed, it answers the question why she failed given that she had three children and twenty-three strawberries—and this question, I have said, is most naturally read as a request, not for a reason why she failed but as a request for reasons why the facts given are reasons why she failed.

I just said that that "Because distributing them evenly was impossible" does not answer the question why Jane failed, and what I meant was that it does not answer the question why Jane failed to distribute her straw-berries evenly among her children. One might object that it *does* answer the question why she failed to distribute her *twenty-three* strawberries among her *three* children. But I think that if the proposition expressed by "Because distributing them evenly was impossible" is indeed an answer to this question, this question *just is* the question-with-presupposition of why Jane failed given that she had three children and twenty-three strawberries. When one asks a question-with-presupposition, the presup-position need not be explicitly flagged by a "given that," or left completely implicit; it may appear in the interrogative in a different way.

Another example: I point at a tree, and ask why that apple tree is putting out apples. You answer: because it is in the nature of apple trees to put out apples. That is a good answer. But the fact that it is in the nature of apple trees to put out apples does not cause this apple tree to put out apples. Counterexample? No. When "because it is in the nature of apple trees to put out apples" is the answer, the interrogative is being interpreted as a question-with-presupposition, namely the question of why, given that this tree is an apple tree, it is putting out apples. The cause (the relevant cause in this context) of its putting out apples is the fact that it is an apple tree; the fact that it is in the nature of apple trees to put out apples is a second-level reason why this cause is a reason.

Think about another why-question one might ask about the tree: why is *this* tree putting out apples? The response might be: because it is an apple tree, and it is in the nature of apple trees to put out apples. This is structurally just like the response in the SECOND ROCK DIALOGUE:

B: It landed at a speed of 4.4 m/s because I dropped it from 1 meter, and Newton's theory of gravitation entails that, for short falls, the impact speed s is related to the distance fallen d by the equation $s = \sqrt{2dg}$, where g is the gravitational acceleration near the surface of the earth. And of course $\sqrt{2 \cdot 1 \cdot 9.8} \approx 4.4$.

The difference is that in the response to the question about the tree the second conjunct is a claim about the nature of apple trees rather than a law of physics. As one might expect, my view about this response is the same as my view about the response in SECOND ROCK DIALOGUE: the first conjunct, namely that it is an apple tree, reports a reason why the tree is putting out apples; the second conjunct is not part of the answer, it instead is a reason why the first conjunct is a reason. When someone asks why this apple tree is putting out apples, he presumably already knows that one reason why the tree is putting out apples is that it is an apple tree; so, once again, the natural interpretation of his question has him asking why, given that it is an apple tree, it is putting out apples.

I said two paragraphs back that, with respect to STRAWBERRIES and BRIDGES, "because it cannot be done" offers a second-level reason, a reason why the causes of Jane's and Jones's failures are causes. But I do not think that this is always what is going on when we answer a why-question with "because it cannot be done." "Fort Knox has not been broken into because it cannot be broken into"—this sounds fine. I take it no one thinks this is a "non-causal explanation," even though the fact that it is impossible to break in to Fort Knox is not a cause of anything, and so not a cause of anyone's failure to break in. I think what is going on in this case is "because it cannot be done" is being used to get across the claim that Fort Knox's defenses are sufficient causes of each person's failure to break in, in the sense that no other causes were needed to bring about that failure. (This is not to say that no failure had any other causes, just that those other causes were not necessary for the effect.) This is another possible interpretation of "because Jane had to fail," said about STRAWBERRIES. The point of that response could be that the facts that she had three children and twenty-three strawberries were sufficient causes; no other causes were needed, perhaps there were no other (relevant) causes.[17]

---

[17] One might say that Jane's birth was a cause of her failure, but clearly it is not a relevant cause in this context.

One more thing Lange says about the Königsberg example is worth discussing. Suppose that Jones tries again and again to cross all the bridges exactly once. Suppose every resident of Königsberg tries. All of Immanuel Kant's descendants. They all fail, every time. Lange says that his answer to the question why Jones failed is an answer to the question why everyone failed. "[T]he distinctively mathematical explanation of the repeated failure to cross the Königsberg bridges," he writes,

> shows that it cannot be done (where this impossibility is stronger than physical impossibility) and so that it was *no coincidence* that all actual attempts failed.   (pp. 505–6; my emphasis)

The idea is that a collection of "merely causal explanations" of each of the failures would not show that it was no coincidence that everyone failed.

I disagree. True, if I said: Jones failed on his first attempt because he ran out of time and had to leave for an appointment; he failed on his second attempt because he gave up partway through in frustration; and so on— and if all of these were true, then yes, for each of his attempts, I would have answered the question why he failed on that attempt, but for all that I would have said, the fact that he failed every time would be a coincidence. But these answers to the questions why he failed are not complete. There is a fact that does not appear in them, a fact that does appear in each complete answer: the fact that the bridge set-up lacks P.

So suppose instead that I said this in response to the question why Jones failed on his various attempts: one reason why Jones failed in his first attempt was that the bridge set-up lacked P; one reason why he failed on his second attempt was that the bridge set-up lacked P; and so on. This is a collection of "merely causal explanations." But they show that it was no coincidence that he always failed. For to show that it is no coincidence that all the facts in some collection obtain it is enough to find a *common reason* why they all obtain—which is what my answers do.

5.4 A Few More Examples

What I've said about why-questions with presuppositions provides a different perspective on equilibrium explanations. Take a simpler example than Fisher's sex-ratio example: we find a marble at rest in the center of the bottom of a bowl. The question is why. My view is that the only relevant reason why it is at rest at the center is that it has been inside the bowl for a sufficient amount of time (and didn't start out moving so fast that it shot right out of the bowl). But in most contexts, anyone asking why the marble is at rest in the center already knows the marble has been there for a while, and so already knows, of the only relevant reason why the marble is at rest, that it is a reason. So in most contexts the why-question is a question-with-presupposition: given that the marble has been in the bowl for a while, why is it at rest in the center? And it is natural to interpret this question as a request for reasons why the presupposed reason is a reason. That is why, I think, so-called equilibrium explanations go about showing that the state the system is in is an equilibrium state, instead of describing the causes of the system's being in that equilibrium state.

Jane failed to distribute her strawberries evenly among her children, but she could have succeeded, if only she'd had a different number of children or strawberries. What about cases where someone fails at something, but could not have succeeded? Smith's mathematics teacher sets a prize for anyone who can trisect a 60 degree angle. Smith tries and tries, filling up pages of notebooks, but fails every time. Why? Surely the answer is: because it is *mathematically impossible*.

What I have to say about this example is, for the most part, predictable from what I have said about others. But there is one tricky thing about it. There are two different readings of the interrogative "Why did Smith fail to trisect a 60 degree angle?" On one reading it is equivalent to "Why did Smith fail, rather than succeed?" where this means "Why did Smith fail, when he could have succeeded?" This question has no answer, because it has a false presupposition, namely that Smith could have succeeded. If someone asks this question, a good thing to say back is that success was mathematically impossible. But by saying this, one does not answer the question why he failed, one rejects it as based on a mistake.

There is, though, another, far less natural, reading of "Why did Smith fail?" That reading makes it equivalent to "Why did Smith fail, when his failure did not have to occur?" This question does not have a false presupposition; although Smith could not have succeeded in trisecting the angle, his failure did not have to occur. His failure would not have occurred if he had not existed at all (or not at the time when that failure occurred). The fact that Smith existed, then, seems to be a, and maybe the only relevant, cause of his failure.[18] The rest will be familiar: if "because success was mathematically impossible" is heard as answering the question why Smith failed (rather than heard as a rejection of the question), then the question being answered is the question why Smith failed, given that he existed, even though his failure needn't have occurred; and what the answer does is provide a reason why his existence is a cause of his failure, and a reason why he failed.

It might be thought that events that happen of necessity are a problem for my theory, since it might be thought that if an event happens of necessity, the fact that it happens of necessity is the reason why it happens. But if a particular event's occurring is necessary, it is either conditionally necessary, or unconditionally necessary. Jane's failure to distribute her strawberries evenly was not unconditionally necessary; it was only necessary given facts about how many strawberries and how many children she had. Events that are only conditionally necessary do not worry me, for I think that the conditions are causes of the event, and that the necessity is not a reason why the event occurs, but instead is a reason why the conditions are its causes. What about unconditionally necessary events? Are they a problem for my view? It is harder to find examples of unconditionally necessary events than it might at first appear. Smith's failure to trisect a 60 degree angle might at first have looked like it was unconditionally necessary, but it wasn't. There were just fewer conditions on its occurrence than on Jane's failure. The same goes for the event of a massive body's moving at less than the speed of light. This is not unconditionally necessary, given the laws of special relativity. It is conditioned on the facts that (i) the body exists at all, and (ii) has a non-zero mass. The laws of special relativity do not constitute a reason why the body is moving at less than the speed of light; what they constitute is a reason why its existing, and having non-zero mass, are the (only) causes of its moving at less than the speed of light.[19]

---

[18] Maybe the fact that the big bang happened is also a cause of his failure, but it is not an interestingly different cause in this context.

I do not think that trying is a precondition for failing, so do not regard Smith's trying as a cause of his failure. While we usually presuppose that X tried when we say that X failed, the second claim does not entail the first. Everyone is always failing to trisect a 60 degree angle, whether they are trying to or not. But it doesn't matter whether you agree; what I say about the case can be easily modified to accommodate the conviction that his trying is one of the causes of his failing.

[19] I will say the same about Mark Colyvan's "non-causal explanation" of the fact that, right now, there are antipodal points on the earth with the same pressure and temperature

So what about events that truly are unconditionally necessary? Could they be counterexamples to my view? I'm not sure I believe that there are (or could be) any such events. The most straightforward way to try to establish their existence is to suggest that the event of its either raining or not raining right now is one. But although I am a liberal about what events there are, I am not that much of a liberal.

In the face of examples of events where some law of nature, or of mathematics, appears to be a reason why those events happened, I keep saying that those laws are instead reasons why the causes of those events are reasons. This might inspire someone to look for *uncaused* events where some law of nature, or of mathematics, appears to be a reason why those events happened. My strategy will not work for them. In fact my strategy will not work on any case in which (i) some event has no causes, but (ii) some fact or other (law or not) appears to be a reason why that event happened (without being a ground of its happening). Are there examples like this?

Reasons why some uncaused event happened have come up before, when I discussed stellar collapse, and Lewis's discussion of stellar collapse. The stellar collapse example was supposed to be an example where the fact that a certain star stopped collapsing was uncaused, yet something, maybe the Pauli Exclusion Principle, was a reason why it stopped collapsing. I already said that this example doesn't work the way it should. But that needn't stop us from finding another example that works the way this one is supposed to.

Lewis alludes to one ("Causal Explanation," p. 222); more fully developed, it goes like this. Space has a boundary, an edge. There is nothing beyond the boundary, not even empty space. A ball rolls toward the boundary, gets to the boundary, and then stops. Why did it stop? Nothing caused it to stop;[20] it stopped because it got to the edge of space. But this fact about the "shape of space" is not a cause of its stopping; so (T1) is false.

(*The Indispensability of Mathematics*, p. 49). This way of dealing with these examples differs from my way of dealing with them in "Are There Non-Causal Explanations?", where I held (i) that these events could not be prevented, and (ii) saying that they could not be prevented constituted a causal explanation.

[20] I don't really think that this event is uncaused (nor does Lewis). That the ball existed earlier is a cause of its stopping. But it is not plausible to say that the fact that the ball reached the edge of space is (only) a reason why its having existed earlier is a cause of it stopping. So I am ignoring this kind of cause.

I do not accept this. Why isn't the fact that it got to the edge a cause of the fact that it stopped? If space had extended beyond the edge (this is a possible state of affairs), then the ball wouldn't have stopped. I'm willing to count the fact that space does not extend past the boundary as a cause of the stopping.

Maybe not everyone is willing to go along with this. But if one is not willing to count the fact that there was no space beyond the edge as a cause of the stopping, one should not, I claim, be willing to say that the fact that there was no space beyond the edge was a reason why the ball stopped.

Anyway, there's a way of thinking about the example on which the fact that the ball got to the edge is not a reason why it stopped but is merely a higher-level reason why something else is a reason why it stopped. It is not metaphysically necessary that things stop when they get to the edge of space. Here's something else they could do: vanish, and appear on the "opposite side" of space, moving in the same direction as when they vanished (as Ms Pac-Man does when she gets to the edge of the screen). Given that the ball could just vanish and go somewhere else when it gets to the edge, the fact that there is no space beyond the edge cannot really be the reason why it stopped. There must be some forces, coming from somewhere, with the feature that they slow down and then stop things as they get close to the edge of space. If there are forces like that, then they, and not the geometry, are the reasons why the ball stops. If the geometry has anything to do with why the ball stopped, it is just a reason why the force laws are such as to put forces of that kind near the edge of space.

The edge of space example was supposed to be an example of an event with no causes where there were still reasons why it happened (other than grounds). The example turns out to be more complicated than it looks at first, and to not be an example of the kind wanted. It is much harder than one might have thought to come up with examples of that kind.

I of course think it is impossible to come up with examples of that kind; but we should at least try a few more. For I imagine my opponents might say that it is easy to construct such cases if we are willing to consider possible worlds governed by very strange laws of nature. Suppose that it is a law that every ten minutes an electron appears at some point of space X. Lo, an electron e appears at X. Isn't this true: there are no causes of e's appearance; nevertheless, the reason why e appeared at X is that it is a law that an electron appears at X every ten minutes.

I don't think this is true. Electron e's appearance *is* caused: it is caused by the fact that an electron appeared at X ten minutes ago. (Maybe this is better put using dates rather than the modifier "ten minutes ago": if e appeared at X at time T, then the cause of its appearance is the fact that an electron appeared at X at time T–10.) The law, again, is a (higher-level) reason why this fact is a cause, not a (first-level) reason why e appeared at X.

My response to this example might inspire consideration of even stranger worlds governed by stranger laws of nature. What if it is a law that an electron appears at X *at time T*? Lo, an electron appears at X at T. Surely now its appearance is uncaused, but still happens for a reason: the reason why it appeared is that it is a law that an electron appears at X at T.

At this point I am going to refuse to continue to play along. I don't see any reason to take the possibility of laws like this seriously. I don't see any reason to regard this scenario as a possible scenario. (The idea that laws of nature, or at least laws of physics, cannot make "direct reference to individuals" is common enough.)

There are yet other examples where a law seems to be a reason why some event happened, that don't turn on the event's being uncaused. This one is due to Hempel: I light a powder I have on fire, and it burns with a yellow flame. Why? Because the powder is a sodium salt, and it is a law that all sodium salts burn with a yellow flame.[21] Isn't the fact that it is a law that all sodium salts burn with a yellow flame a reason why this powder is burning with a yellow flame?

Does my usual strategy work with this example? Can I say: the law is a higher-level reason why some cause is a cause? The cause would have to be the fact that this powder is a sodium salt. This doesn't sound to me like a bad thing to say. After all, if the powder had not been a sodium salt, it would not have burned with a yellow flame.

I also have a different response to this example that doesn't depend on the fact that the powder's being a sodium salt is a cause. I still say that "one reason why this powder burned with a yellow flame is that it is a law that all sodium salts burn with a yellow flame" is false. Why does it nevertheless sound good to mention the law in response to the why-question? Because the law constitutes a partial answer to the question

[21] "Deductive-Nomological vs. Statistical Explanation," p. 125.

why the powder burned yellow. The law tells us something about the reasons why the powder burned yellow, without identifying any of those reasons in particular. What does it tell us? That the reasons why it burned yellow are also reasons why every other sodium salt burned yellow. More carefully, the law tells us that there is a kind of reason with the property that, whenever a sodium salt is burning yellow, a reason of that kind is a reason why it is burning yellow.

The law doesn't tell us what kind of reason this is, but for the record: the reason why this powder burned yellow is that during the burning process the electrons in some of its constituent sodium atoms "de-excited," and the difference between those electrons' initial and final energies (the energies of the 3p and 3s states respectively) corresponds to the energy of photons with a wavelength of roughly 589 nanometers (light with this wavelength is yellow). So the kind of reason we're after can be characterized like this: a reason is of this kind if it differs from this one only with regard to which powder it concerns.

What I've said about the burning salt seems like the right thing to say about plenty of other cases too. Why is this thing black? It is a good response to say: well, it's a raven, and all ravens are black. And that sounds like a way of saying: I don't know the reasons why it's black, but I can tell you this, they are just the same as the reasons why each other (normal) raven is black.[22]

---

[22] Albino ravens that have been painted black are not black for the same reasons that normal ravens are.

6

Teleology and Reasons for Action

6.1 Teleology

Since chapter 2 I have restricted my attention to answers to why-questions about concrete events. In addition to that restriction I made two more: I set aside teleological answers to why-questions; and I set aside the case where the event asked about is an intentional action, and the why-question is a request for the agent's reasons for acting.

I think that my theory is interesting enough, and controversial enough, even as a theory of this restricted domain. Still, I do want to say something about whether, and if so how, my theory can be extended to a theory of answers to why-questions about concrete events generally, without the restrictions.

First I will lift the first restriction, and consider teleological answers to why-questions that are not requests for an agent's reasons for acting. I want to say up front that my thoughts about this kind of answer are not as well-developed as my thoughts about non-teleological answers. As a consequence, my discussion here is more tentative and exploratory than it was in the earlier chapters.

In this section I am going to state four different theories of teleological answers. After the first one, each theory is motivated by some problem I see with the one that came before. But I am not entirely convinced that I end up in the right place.

Since chapter 2 I have been using " 'in order to' answers" as shorthand for teleological answers to why-questions. Many philosophers have thought that "in order to" answers can be analyzed in terms of answers that cite causes. At night, a plant might close its stomata[1] in order to

---

[1] Stomata are pores in the "skin" of a plant's leaves.

conserve water. The general strategy is to say that "the plant closed its stomata in order to conserve water" is true iff the plant was caused to close its stomata in a certain sort of way; the challenge is to figure out what kind of causes make "in order to" answers true.

It might seem like defending my theory of reasons why requires me to sign on to this research program. This is not true. But the argument for this conclusion is not as straightforward as it may seem at first.

At first the following argument might seem like a good one: the core of my theory is a theory of reasons why. But, one might think, answers of the form "X ϕ-ed in order to Z" do not report reasons why. If I want to have a complete theory of answers to why-questions, I will of course want a theory of "in order to" answers. And it is certainly in the spirit of my theory of reasons why to want a causal theory of "in order to" answers. But my theory of reasons why does not "commit me" to a causal theory of "in order to" answers.

While I accept this conclusion, this argument rests on a false presupposition: that teleological answers to why-questions do not report reasons why. In fact they do. "X ϕ-ed in order to Z" is equivalent to "The (or a) reason why X ϕ-ed was to Z." "The plant closed its stomata in order to conserve water," "the reason why the plant closed its stomata was to conserve water"—these "say the same thing," in some sense.

The observation that "in order to" answers do provide reasons why might suggest that I *am* committed to a causal theory of "in order to" answers. The observation might further suggest that my theory of reasons why, (T1f), is false; for if "The reason why the plant closed its stomata was to conserve water" is true, we seem to have a reason why that is neither a cause nor a ground. But these suggestions are not right. Yes, the reason why the plant closed its stomata that this answer offers is not a cause of the stomata's closing. Yes, I often summarize my theory as "reasons why are causes or grounds." But officially the sentence-form my theory is a theory of is "the reason why Q is that R"—and the sentence giving the reason why the plant closed its stomata is not of this form.

Throughout this book I have taken reasons to be facts (which in turn may be taken to be true propositions). Facts have contents that can be given by a that-clause. Each fact is the fact that such-and-such. It is these "factual reasons why" that I have proposed a theory of. But in "The reason why the plant closed its stomata was to conserve water" the clause that gives the "content" of the reason is not a that-clause, but is instead a non-finite clause starting with "to." "To conserve water" is not, and cannot be, a fact. So this reason why the plant closed its stomata is not a factual reason why; its failure to be a cause of the closing, therefore, is not a problem for my view.

My causal/grounding theory of factual reasons why does not commit me to a causal theory of what we might call "non-finite" reasons why (taking their name from a grammatical feature of the clauses used to report them). The failure of every causal theory of non-finite reasons why would not show my theory of factual reasons why to be false. For all that I have argued so far, I could uphold the autonomy of teleological answers to why-questions. Still, as I said, it is certainly in the spirit of my theory to want a causal theory of non-finite reasons why.

Here is a route to one promising causal theory of non-finite reasons why. We start with the idea that what it takes for some behavior to be for the sake of a certain goal or end is for that goal or end to play a role in producing—causing—that behavior.[2] But how can a goal or end help cause a piece of behavior? The bit of "The reason why the plant closed its stomata was to conserve water" that has to do with the goal is the phrase "to conserve water." But this has the wrong grammatical properties to describe a cause.[3] It makes no sense to say that the plant was caused to close its stomata by "to conserve water." So we need to find a way to build

---

[2] Behavior that is done for the sake of a goal is "goal-directed." Woodfield (*Teleology*, pp. 42–3) notes that "goal-directed" is ambiguous: "It could mean 'directed to a goal', rather as 'homeward-bound' means 'bound for home'; or it could mean 'directed by a goal', in the way that 'handmade' means 'made by hand'. On the second interpretation the goal causally influences the direction . . ." A causal theory of non-finite reasons why starts with the idea that behavior that is goal-directed in the first sense is so in virtue of being goal-directed in something like the second sense (not exactly the second sense, since as I say in the main text goals cannot be causes).

[3] Similarly, the bit of "I am running my hand over the table in the dark in order to find my keys" that has to do with the goal is "to find my keys." In this case we can extract from the goal a "goal-object," namely my keys. My keys (or facts about them) are certainly capable of causing things, such as the starting of a car. Goal-objects can cause things, even if goals cannot. This suggests an hypothesis: the reason why X ϕ-ed was to Z iff the goal-object associated with Z-ing played some role in causing X to ϕ. But this is a dead end. An instance of "The reason why X ϕ-ed was to Z" can be true without there being any natural way to extract a goal-object from "to Z"—what was the goal-object when the reason why some animal exhaled was to rid itself of excess carbon dioxide? (Extending an objection from Taylor's "Purposeful and Non-Purposeful Behavior," Scheffler raised "the difficulty of the missing goal-object" for some theories of teleology in his 1959 paper "Thoughts on Teleology"—see p. 268.)

reference to the goal or end into something that is at least a potential cause of the behavior. Here is a suggestion: *the fact that closing its stomata was a way for the plant to conserve water* both (i) contains reference to the goal and its relation to the behavior, and (ii) is the right sort of thing to be a cause of that behavior (since it is a fact). The theory we get by requiring it to cause the behavior looks like this:

> TELEOLOGY-1: The reason why X ϕ-ed was to Z iff the fact that ϕ-ing was a way (or part of a way) for X to Z caused X to ϕ.

I said that it is "in the spirit" of my theory of factual reasons why to want a causal theory of non-finite reasons why. Partly I just meant that then my theories of both kinds of reasons would centrally involve causation. But there is more: if TELEOLOGY-1, or someting like it, is true, then non-finite reasons why "reduce" to factual reasons why. For TELEOLOGY-1, in conjunction with (T1f), is almost the theory that the reason why X ϕ-ed was to Z iff the reason why X ϕ-ed was that ϕ-ing was (part of) a way to X to Z.[4]

The object of analysis in TELEOLOGY-1 is the locution "The reason why X ϕ-ed was to Z," which I have been treating as equivalent to "X ϕ-ed in order to Z." So TELEOLOGY-1 is not, by itself, a comprehensive theory of teleological answers to why-questions. A comprehensive theory must provide truth-conditions for every form of words that can be used to supply a purpose in response to a why-question. There is a great variety of forms that can be used to do this. (I listed some of them in footnote 26 in chapter 2.) I will not, however, discuss the truth-conditions for those other forms in detail, just as I did not discuss in detail what it takes for sentences of the form "Q because R" to be true when I presented (T0) and (T1). I do assume that those other forms can be analyzed in terms of non-finite reasons why, or, if not, that there is a natural adaptation of TELEOLOGY-1 to cover them.[5]

---

[4] "Almost" because the right-hand side could be true if the fact that ϕ-ing was (part of) a way for X to Z was a *ground* of the fact that X ϕ-ed, while the right-hand side of TELEOLOGY-1 could not. If it could never happen that the fact that ϕ-ing was (part of) a way to X to Z grounded the fact that X ϕ-ed, then TELEOLOGY-1 is equivalent to the stated theory, given (T1f).

[5] It may be worth saying something about one other form in a footnote. "Jane sent Marge to Texas to meet Ross" can be the answer to the question why Jane sent Marge to Texas, but it is not equivalent to "Jane sent Marge to Texas in order to meet Ross." In the first sentence the goal is for Marge to meet Ross, in the second, for Jane to meet Ross (the sentence suggests that sending Marge to Texas is a way for Jane to meet Ross—see Huddleson and Pullum,

Many have complained that, since teleological answers to "Why did X φ?" refer to the goal at which the φ-ing is aimed, and since the φ-ing precedes the attainment of the goal at which it is aimed, teleological answers can be true only if there is backwards causation, which there is not.[6] If the complaint is that teleological explanations are true only if goals cause behavior, then it is off-base; teleological explanations do not say that goals cause behavior—goals, characterized by to-infinitival clauses, are not the right sort of thing to cause anything.

Of course, *the fact that X attained its goal* is the right sort of thing to cause X to φ. If you think that the only way to get reference to the goal into a cause of X's φ-ing is to make this fact a cause, and so if you think that a causal theory of non-finite reasons why has to start like this

(1)   The reason why X φ-ed was to Z iff (i) the fact that X (later) Z-ed caused X to φ, and (ii) . . . ,

then the "no backwards causation!" objection will sound good to you. But (1) is not the only way to get reference to the goal into a cause of X's φ-ing.

TELEOLOGY-1 is my own proposal, though it is close to Larry Wright's theory.[7] Wright's theory is less specific than TELEOLOGY-1: he says that X φ-ed in order to Z iff φ-ing "is behavior that occurs because it brings

Cambridge Grammar, pp. 728–9). But "Jane sent Marge to Texas in order to meet Ross" *is*, I think, equivalent to "Jane sent Marge to Texas in order to make it the case that Marge meets Ross," so TELEOLOGY-1, as stated, can be made to apply to "Jane sent Marge to Texas in order to meet Ross."

The Jane-Marge-Ross example raises an issue: I have been treating "The reason why Jane sent Marge to Texas was to meet Ross" as equivalent to "Jane sent Marge to Texas in order to meet Ross"; but now we have an alternative hypothesis, that it is equivalent instead to "Jane sent Marge to Texas to meet Ross." Which hypothesis is right? I hear "The reason why Jane sent Marge to Texas was to meet Ross" as having two readings, one equivalent to "Jane sent Marge to Texas in order to meet Ross," the other equivalent to "Jane sent Marge to Texas to meet Ross." I should disambiguate: "The reason why X φ-ed was to Z," as it appears in TELEOLOGY-1, is meant to be interpreted so that (the term that goes in for) "X" is the (missing) subject of the infinitival "to Z." This has the effect of making my earlier assumption true: that ". . . reason why is to . . ." statements are equivalent to "in order to" statements.

[6] See Wright, *Teleological Explanations*, pp. 10–11, for one discussion of this complaint. It goes way back: see Des Chene, *Physiologia*, p. 188, for a discussion of it in the context of late Aristotelian thought.

[7] Wright, *Teleological Explanations*. His theory builds on Taylor's theory in *The Explanation of Behavior*. Cohen later defended a similar theory (*Karl Marx's Theory of History*, chapter 9; the theory is too influenced by Hempel's view that explanations must cite laws). Wright's theory seems to have made earlier theories obsolete, so I will not survey those theories; see Scheffler, "Thoughts on Teleology," Wright, "The Case against Teleological

about, is the type of thing that brings about, tends to bring about, is required to bring about, or is in some other way appropriate for bringing about" Z (pp. 38–9). I want to focus on a proposal that is less open-ended and disjunctive.

In another way Wright's theory seems to me too restrictive. He requires the relationship between the ϕ-ing and Z-ing (the behavior and the goal) to be causal. The behavior has to "bring about" (or tend to bring about . . .) the goal. As far as I understand what he means by "bring about," he means something causal. The behavior must cause the achievement of the goal (or "tend to" cause this). This works fine for the main example I have been using. By closing its stomata, the plant caused water to be conserved. But behavior and goal need not be related as cause to effect. When I watch television in order to be doing something, watching TV does not cause me to be doing something. It "realizes" my doing something. Speaking of "ways" to Z, as TELEOLOGY-1 does, is a way to be neutral between causing and realizing a goal. Even when watching TV does not cause me to be doing something, it is a way to be doing something.[8]

Even though TELEOLOGY-1 does not require the behavior to cause the achievement of the goal, TELEOLOGY-1 is still a causal theory; while the behavior need not cause the achievement of the goal, *the fact that the behavior is a way to achieve the goal* must cause the behavior.

The bit about "part of a way" in TELEOLOGY-1 is there for the following kind of case: suppose I drop a rock and it falls three feet to the ground. To fall three feet it first fell one foot, and then two more. If I believed that rocks could exhibit goal-directed behavior, and I believed that the end rocks aimed at was being on the ground, I would want to say that the rock fell straight down for one foot in order to get to the ground. But falling straight down for one foot is not a way to get to the ground, it is only part of a way. (In what follows I will often ignore this "part of" qualification.)

Mark Bedau proposed this counterexample to theories like TELEOLOGY-1:

Consider a stick floating down a stream which brushes against a rock and comes to be pinned there by the backwash it creates.

Reductionism," Woodfield, *Teleology*, Parts 2 through 4, and Bedau, "Where's the Good in Teleology?", for surveys and criticisms. (I will come to Bedau's criticism of Wright shortly.)

[8] Woodfield suggests a similar amendment to Wright's view (*Teleology*, p. 87).

The stick is creating the backwash because of a number of factors, including the flow of the water, the shape and mass of the stick, etc., but part of the explanation of why it creates the backwash is that the stick is pinned in a certain way on the rock by the water. Why is it pinned in that way? The stick originally became pinned there accidentally, and it remained pinned there because that way of being pinned is self-perpetuating. Therefore, once pinned, part of the explanation for why the stick is creating the backwash is that the backwash keeps it pinned there and being pinned there causes the backwash. . . . creating the backwash tends to pin the stick on the rock and the stick creates the backwash because doing so contributes to pinning it. Clearly, however, the stick does not create the backwash *in order to* keep itself pinned on the rock.

> ("Can Biological Teleology Be Naturalized?", p. 648;
> see also "Where's the Good in Teleology," p. 786.)

I just don't see it. I don't see why I should accept that the fact that creating the backwash is a way to pin the stick to the rock causes the stick to create the backwash. The sense in which the stick's situation is self-perpetuating seems to be this: its being pinned causes it to create the backwash, and its creating the backwash causes it to be (or remain) pinned. This does not seem to me to add up to *the fact that the backwash is a way to pin the stick* causing the stick to remain pinned.

Some theories of teleological explanation require the goal to be in some way good. In the context of TELEOLOGY-1, the claim would be that TELEOLOGY-1 should be made more demanding, and say that the reason why X ϕ-ed was to Z iff (i) the fact that ϕ-ing was a way for X to Z caused X to ϕ, and (ii) Z-ing was in some way good.[9] I am not persuaded. Insofar as I can imagine a world in which rocks move down in order to get to the center of the universe, I can imagine one in which rocks do

___

[9] Woodfield advocates this view in *Teleology* (see p. 205 and following), as does Bedau in, for example, "Where's the Good in Teleology." Cooper, in "Aristotle on Natural Teleology," says that it was Aristotle's view that goals are in some way good (for example on p. 107). An analogous thesis about functions is perhaps more plausible, that thesis being that it cannot be the function of X to Z unless Z-ing is in some way good, perhaps for some "system" X is part of. (See Manning, "Biological Function, Selection, and Reduction"; thanks here to Kieran Setiya. I will say something brief about functions at the end of this section.)

this but there is nothing good (or bad) about getting to the center of the universe.[10]

There are nevertheless problems for TELEOLOGY-1. First and foremost is the question of how the fact that ϕ-ing was a way for X to Z could ever have managed to cause X to ϕ.

I think that it is certainly possible, in some broad sense of "possible," for the fact that ϕ-ing was a way for X to Z to have caused X to ϕ. I can imagine a scenario in which the usual counterfactual tests for causation come up positive. I drop a rock and it moves down; if moving down had not been a way to get to the center of the universe (if the center had been located in a different direction from the rock's position), the rock would not have moved down; if moving in direction D had instead been the way to get to the center of the universe, the rock would have moved in direction D. For these counterfactuals to be true, the laws of physics would have to be very different; but I don't see any obstacle to there being laws that make these counterfactuals true.

You might object that the truth of these counterfactuals does not really help; the real, more basic, problem is that the fact in question, the fact, call it F, that moving down is a way for the rock to get to the center of the universe, couldn't be a cause of *anything,* no matter what the circumstances. It is relevantly like the fact that $2 + 2 = 4$, which also couldn't be a cause of anything. One might put the point this way: only facts that correspond to concrete events are eligible to be causes, and neither F nor the fact that $2 + 2 = 4$ corresponds to a concrete event. In reply, I will say that I don't think that F is relevantly like the fact that $2 + 2 = 4$. I do think it corresponds to a concrete event. Maybe it helps to note that, given that rocks get from place to place by moving continuously in space, F is

----

[10] The stick/backwash example, discussed above, is one of Bedau's arguments; his diagnosis of the example is that it meets the causal requirements that people like me say are sufficient for its being the case that the stick creates the backwash in order to pin itself to the rock; since in fact the stick does not create the backwash in order to pin itself, Bedau suggests that what is missing, what is needed for pinning itself to be the goal of creating the backwash, is for there to be something good about the stick's pinning itself to the rock. As I said above, I dispute Bedau's claim that the example meets the causal conditions.

When we are discussing intentional actions, rather than the movements of rocks, the idea that there must be something good in the goal, for someone to have acted in order to achieve that goal, is more plausible. For G-ing to be the goal of an intentional action of mine, mustn't I want to G? And if I want to G, mustn't I see G-ing as in some way good? Kieran Setiya argues that this is false in *Reasons Without Rationalism* and in "Sympathy for the Devil."

equivalent to the fact that the center of the earth is located at the center of the universe; and this fact obviously corresponds to a concrete event, obviously can be a cause.[11]

Even if we accept that F corresponds to a concrete event, there is another problem with what I said two paragraphs back, when I said that F passes a counterfactual test for causation. There are possible laws that would make those counterfactuals true, but would also make the outcome of using those counterfactuals to test for causation a "false positive." Consider for example a world in which Newton's second law is true, but in which the "source laws" for forces are not what we are used to. Instead of a law for gravitational forces, and one for electromagnetic forces, and so on, there is just one law: the force on a body is always directed toward the center of the universe (the magnitude of the force isn't important). Rocks in this universe always move toward the center of the universe, but they do not move toward the center in order to get to the center. At least that sounds to me like the right thing to say. When a rock moves toward the center of the universe under these laws, the fact that the force acting on the rock was directed toward the center of the universe is a cause of the rock's motion; the fact that moving in that direction was a way to get to the center does not seem to be a cause. Teleological laws, laws that make it true that the rock was caused to move by the fact that moving like that was a way to get to the center of the universe, have to entail in some other way (not via Newton's second law) that bodies (or bodies made of "earth") always move toward the center of the universe.

What might such laws look like? In their paper "What Would Teleological Causation Be?" Hawthorne and Nolan state some possible laws and claim that, were they laws, they would be teleological laws. Simplifying a little, one of their examples is this: for anything made of earth (like a rock) and any time T, of all the ways for that thing to start from its location at T and go to the center of the universe, at T it begins to take the fastest of those ways.[12] It seems right to me that, if that were the law, and a rock

---

[11] I'm assuming here that "down" always denotes the direction toward the center of the earth; if it is taken, in the example, to denote the direction toward the center of the universe, I would have to put the point differently.

[12] As is well-known, the idea that even rocks exhibit goal-directed behavior is associated with Aristotle. The scenario I just described could be made even more Aristotelian if we said that the law was not a contingent law of nature, but instead something that flowed from the nature of things made of earth.

began to move in direction D, then the fact that moving in D is the "first step" along the fastest way to the center of the universe does indeed cause the rock to move in D.[13]

I should say, though this is a bit of a tangent, that possible worlds in which rocks obey teleological laws do raise a problem for TELEOLOGY-1, the problem of "ineffective means." Insofar as we accept that a rock could move down in order to get to the center of the universe, we should admit that a rock could move down in order to get to the center of the universe, even when there are obstacles—a table, the earth itself—in the way. But when the earth intervenes between the rock and the center of the universe, moving down is not in fact a way to get to the center of the universe (supposing the center of the universe to fall inside the earth, there are no ways to get to the center of the universe), and TELEOLOGY-1 is false. The solution is to weaken TELEOLOGY-1, and require only that moving down be a way to get to the center of the universe under idealized conditions, where the idealization involves removing obstacles. I won't try to figure out how to make this precise.[14]

Are there cases where something X ϕ-s in order to Z, yet even under idealized conditions, ϕ-ing is not a way to Z? Are there means that are not just ineffective but completely misguided? I suppose I might read Plato's dialogues in order to become a faster runner. But no amount of "removing obstacles" will make for a situation in which reading Plato's dialogues is part of a way to become a faster runner. I am going to say that examples like this do not count right now. TELEOLOGY-1 is not meant to be a theory of non-finite reasons why when those reasons are an agent's reasons for

---

[13] All of Hawthorne and Nolan's examples of teleological laws require things subject to those laws to pursue the "best" way to attain its goal, in some sense of best—"fastest" being only one sense they discuss. I do not think that teleological laws must have this feature. Something can act so as to attain some end without pursuing means that are in any sense best.

Technically, the fact that moving in D is part of *the fastest* way to the center is not identical to the fact that moving in D is part of a way to the center. Given my fondness for the thesis that causes must be proportional to their effects, I ought to say that the first and not the second fact is a cause of the rock's moving in D. So this example does not quite satisfy the right-hand side of TELEOLOGY-1. One possible solution is to weaken TELEOLOGY-1 and require only that the fact that ϕ-ing is a way for X to Z be *part of* a cause of X's ϕ-ing—though I am not sure how to define "part of a cause" so that this gets the right results.

[14] Woodfield also appeals to obstacle-removing idealizations to deal with ineffective means (in a discussion of a different theory of teleological explanation, one he ultimately goes on to reject) (*Teleology*, p. 49).

acting. And I think that you can only get goal-oriented behavior that involves completely misguided means when the behavior is an intentional action.[15] (I should, of course, have something to say about the connection between non-finite reasons why that are not, and non-finite reasons why that are, an agent's reason for acting; and I will, in the next section.)

So far we haven't seen any reason to abandon or modify TELEOLOGY-1; now I want to discuss a problem case that does motivate modifying it. I'm working the remote on my remote-controlled car, causing it to drive around the lot behind my house. I see that a hole in the fence is a way for the car to escape the lot, and I cause the car to drive through the hole. The fact that driving through the hole was a way for the car to escape stands at the beginning of a chain of causes that passed through my mind and then the remote control and led to the car's driving through the hole; but the car did not (did it?) drive through the hole in order to escape.

A natural thought about this example is that the car, while it did move through the hole in the fence, was not in any sense the "agent" of that motion; nothing that happened, as the car moved through the hole, was something that the car did. Instead *I* was the agent, I was in control of the car's movements, in a way that the car itself was not. And "in order to" statements apply to agents not to patients. "X ϕ-ed in order to Z" can only be true if the event corresponding to "X ϕ-ed" is an event consisting in *X's doing something*, is an action of which X is the agent. We could read this requirement back in to TELEOLOGY-1, as an extra clause on the

---

[15] Suppose that I am completely misguided, and think that opening the windows in my house in the winter is a way to heat it up. I build a robot with a built-in calendar and a built-in thermometer, and design it so that whenever it is winter, and the house is below 70 degrees Fahrenheit, the robot opens the windows. I set the robot free in my house, and it goes and opens a window. Did it open the window in order to cool down the house? I'm not inclined myself to say yes. But if the answer is yes, then we have a case of behavior that is (i) not an intentional action (the robot is not sophisticated enough to act for reasons), but (ii) is completely misguided as a means toward its end.

If I did become convinced that the robot opened the window in order to heat up the room, I would call this a case of a "derived end": the robot gets to have heating up the house as the end of its act because of facts about me, in particular, the fact that heating the house is the end that *I* wanted the robot to achieve by opening the window. Then I would restate my claim in the text as: you can only get goal-oriented behavior that involves completely misguided means when the behavior is either an intentional action, or has a derived end. (If derived ends are possible, then I should say that none of the theories in this chapter are meant to apply to them.)

right-hand side. Then TELEOLOGY-1 could be true even if, in this case, the car did not drive through the hole in order to escape the lot.[16]

Do not read this response as more restrictive than it is; the response does not entail that only creatures capable of beliefs and desires, capable of deliberating about what to do, and of doing things for reasons, can do something in order to achieve some end. "Agent," as I used it in the previous paragraph, was not meant to be as loaded as this. I had something more minimal in mind: the car was the agent of the driving, in the relevant sense, iff the driving was something that the car did. In this sense, plenty of things without minds are agents. Plants close their stomata at night; closing their stomata is something those plants do, so those plants are the agents of the closings.[17]

This is not the end of what I want to say about the car example. I'm not happy leaving things here, because I do not have a theory of agency to appeal to, to make good on the claim that the car was not the agent of the driving. So, even though I do believe that TELEOLOGY-1 should be read as requiring the ϕ-ing to be something that X does, I want to try out another thought about the example.

Rocks, in the Aristotelian worlds I imagined earlier, are not just caused to move down by the fact that moving down is the fastest way to get to the center. They have a *general disposition* to do X when doing X is the fastest way to get to the center. The fact that moving down is the fastest way to get to the center causes a rock to move down by causing it to manifest this disposition. The car, on the other hand, does not have any general disposition to do things that are ways to escape. A second suggestion, then, is that it is the car's lacking a disposition like this that blocks the

---

[16] I have put the point like this: the car drove through the hole, but did not drive through the hole in order to escape, because the car was not the agent of the driving. Maybe instead I should deny *that the car drove through the hole*; it can't be true that the car drove through the hole, the idea goes, if the car was not the agent of the driving. That's just part of the meaning of "drive." If I say this, then the right-hand side of TELEOLOGY-1 had better be false when "drive through the hole" goes in for "ϕ." And it seems that it is false: if the car did not drive through the hole, then neither did anything cause it to drive through the hole. If driving is essentially an exercise of agency, so is being caused to drive, for then being caused to drive is being caused to exercise the agency involved in driving.

[17] I learned of the importance of recognizing a notion of agency that applies to things incapable of belief, desire, or reason from Setiya's "Reasons and Causes" (section 2); he cites Alvarez and Hyman, "Agents and Their Actions" (pp. 243–5), Coope, "Aristotle on Action" (p. 134), and Thompson, *Life and Action* (pp. 122–8).

truth of "The car drove through the hole in order to escape." Here is TELEOLOGY-1 amended to require the relevant disposition:

TELEOLOGY-2: The reason why X ϕ-ed was to Z iff the fact that ϕ-ing was (part of) a way for X to Z caused X to ϕ—by causing X to manifest a general disposition to do things that are ways to Z.

When the fact that ϕ-ing was a way for X to Z caused X to ϕ in some other way, not by causing it to manifest the relevant disposition, we might call the chain of causes leading from the fact that ϕ-ing was a way to Z to the ϕ-ing, a "deviant causal chain." The chain that started with the fact that driving through the hole was a way to escape, passed through my beliefs, and my manipulation of the remote control, to the car's driving through the hole, would then be a deviant causal chain. "Deviant causal chain" is a term from action theory, and I will have a lot to say about deviant causal chains in the action-theoretic context in the next section (where an appeal to the manifestation of dispositions will again play an important role[18]).

Before raising the puzzle about the car that led to TELEOLOGY-2 I was asking how the fact that ϕ-ing was a way to Z could ever cause X to ϕ. I said I thought this was possible under very different laws of nature. But is it possible here and now? Is it possible if X is some familiar thing subject to the actual laws of nature? Thinking about this question motivates yet a third theory of teleological answers.

It is certainly possible here and now, in the actual world, for the fact that ϕ-ing was a way to Z to cause X to ϕ. It is possible if ϕ-ing is an intentional action of X's. It could easily happen that (i) I turn left in order to avoid the cliff, and (ii) the fact that turning left is a way for me to avoid the cliff plays a role in my coming to know that turning left is a way to avoid the cliff, and my knowledge plays a role in causing me to turn. But, again, I still have why-questions that ask for an agent's reasons for acting on the back-burner.

Let's think about the example I have been using: the plant closed its stomata to conserve water. Is it true that the fact that closing them was a way for the plant to conserve water caused the plant to close them? It

---

[18] In fact, the idea of moving from TELEOLOGY-1 to TELEOLOGY-2 came from thinking about the solution to the problem of deviant causal chains in the theory of action that I will discuss later.

doesn't seem so. Even if closing them had not been a way for the plant to conserve water, the plant still would have closed them. For the closing mechanism is light-sensitive, and causes the stomata to close at night. Its being nighttime, and its being a time at which closing its stomata is a way for the plant to conserve water,[19] are correlated conditions, but is it clearly the first and not the second that causes the plant to close its stomata. Had closing them not been a way for the plant to conserve water, but had it still been nighttime, the plant still would have closed its stomata. (The truth of these counterfactuals does not prove that the fact that closing its stomata was a way for the plant to conserve water was not a cause; the same pattern of counterfactuals is true when some effect is overdetermined. But this does not seem like a case of overdetermination.)

Either the plant did not, after all, close its stomata in order to conserve water, or TELEOLOGY-1, and TELEOLOGY-2, are false. Which is it? Holding on to TELEOLOGY-1 or TELEOLOGY-2 and concluding that no plant does anything in order to achieve any end is an interesting position.[20] In many places teleological explanations in biology are dismissed on the basis that nothing can act to achieve some end unless it has a mind, or was designed by something with a mind. But dismissing them because they conflict with TELEOLOGY-1 or TELEOLOGY-2 is different. The problem is not that plants lack minds. TELEOLOGY-1 and TELEOLOGY-2 allow that things without minds can have aims; they allow that, if the laws are right, rocks can move down in order to reach the center of the universe. Interestingly, even these naturalistic, causal theories of non-finite reasons rule out teleological explanations in biology.

Supporting the other option, of accepting teleology in biology and rejecting TELEOLOGY-1 and TELEOLOGY-2, is the fact that both biologists and non-biologists are happy to use teleological language to describe the behavior of lower life forms. Biologists also aim to answer questions about

---

[19] This example is not based on any real knowledge I have of botany; it comes from a botany textbook from 1967 (Greulach and Adams, *Plants: An Introduction to Modern Botany*) by way of Wright's *Teleological Explanations* (p. 9). This much extra information about the example might be useful here though: "conserve water" means "prevent the unnecessary loss of water." Closing its stomata is always a way for the plant to prevent loss of water, but only at nighttime is the loss of water unnecessary. During the day it is a necessary side-effect of being open in order to obtain carbon dioxide.

[20] Starting from a theory similar to TELEOLOGY-1, Cohen is led to endorse the analogous conclusion about a very similar example (*Karl Marx's Theory of History*, pp. 268–9).

behavior so described. Wright gives us this great example, from the aptly-titled *The Directiveness of Organic Activities* (E. S. Russell, 1945):

> Experiments by Kepner and Barker have shown that *Microstoma* eats *Hydra* for the sake of its nematocysts rather than for its food value. (Wright p. 50)

There were two hypotheses about *Microstoma*'s aim in eating *Hydra*; these biologists set out to discover which was true; the experiments they did favored one hypothesis over the other.

I am not a philosopher of language, but widespread use of "X ϕ-ed in order to Z" to describe the behavior of lower life forms is at least some reason to think that some of these descriptions are (literally) true.[21]

The general strategy for modifying theories like TELEOLOGY-1 and TELEOLOGY-2 to make "the plant closed its stomata in order to conserve water" come out true is to "appeal to evolutionary history."[22] One way to use this strategy to amend TELEOLOGY-2 yields

> TELEOLOGY-3: The reason why X ϕ-ed was to Z iff (i) the right-hand side of TELEOLOGY-2 is satisfied, or (ii) some of X's ancestors had the trait of being disposed to ϕ in circumstances similar to those X was in; in those circumstances, when X's ancestors ϕ-ed, they (frequently) Z-ed by ϕ-ing; X inherited this disposition from its ancestors; there was selection, in the ancestral environment, for traits (like this disposition) that led X's ancestors to Z; X's ϕ-ing in this case was a manifestation of this disposition.

When clause (ii) but not clause (i) is satisfied, the fact that ϕ-ing was a way for X to Z is not a cause of X's ϕ-ing. This is a problem: a causal theory of goal-directed behavior has to have the goal somehow play a role in causing the behavior; how does that happen when it is clause (ii) that is

---

[21] Compare Neander: "I suppose it is just barely possible, perhaps, that this apparent explanatory power [of teleological explanations in biology] is illusory, based on hangovers from our Creationist past, or due to our mistaking the metaphorical for the literal, when we speak of 'Mother Nature's intentions', 'evolutionary design', and so on. However the thesis that we are persistently irrational in this respect is psychologically implausible in contrast to a theory of functions that shows such explanations to be legitimate" ("The Teleological Notion of 'Function'," p. 457).

[22] There are lots of examples of this strategy in action; for examples, see Woodfield, *Teleology*, p. 208, and Neander, "The Teleological Notion of 'Function' "—though her focus is on the analysis of "The function of X is to Y," not of "The reason why X ϕ-ed was to Z."

satisfied? More generally: I see a clear rationale that starts from the idea that for X to ϕ in order to Z is for X's ϕ-ing to have, as one of its causes, some fact pertaining to Z-ing, and leads to TELEOLOGY-1, which specifies what that fact needs to be. The theory TELEOLOGY-3 may do better than TELEOLOGY-1 at matching our judgments about whether plants and other life forms (that do not act for reasons) ever do things to achieve certain ends; but it would be disappointing if that were all TELEOLOGY-3 had going for it. Is there any independent rationale for it?

One might hold out hope that, when (ii) is satisfied, the fact that ϕ-ing was a way *for X's ancestors* to Z is a cause of X's ϕ-ing; that would certainly make TELEOLOGY-3 resemble the earlier theories more closely. Assuming, as I have done throughout, that counterfactuals are a good test for causation, what do we think of the relevant counterfactual? If ϕ-ing hadn't been a way for X's ancestors to Z (in relevantly similar circumstances), would it have been false that X ϕ-ed on this occasion? Maybe . . . we are assuming that X's ancestors were disposed to ϕ in those circumstances. If ϕ-ing hadn't been a way for them to Z, presumably they wouldn't have Z-ed. (If closing their stomata at night hadn't been a way for the plant's ancestors to conserve water, they still would have closed their stomata at night, but they wouldn't have thereby conserved water. It may help to imagine that closing their stomata at night ceased to be a way to conserve water only after these ancestors already existed.) Given that there was selection for traits that lead to Z-ing, and that this means there was selection against (competing) traits that don't, X's ancestors wouldn't have had much success in reproducing. If the selection pressure were strong enough, X's ancestors would have died out, and so X would not have existed. And of course if X hadn't existed, X would not have ϕ-ed on this occasion.[23] It looks like the counterfactual comes out true. This suggests that, when (ii) is satisfied, the fact that ϕ-ing was a way for X's ancestors to Z *is* a cause of X's ϕ-ing.

Still, even if TELEOLOGY-3 does identify an appropriate cause of X's ϕ-ing, it seems far too specific. When it is clause (ii) that is satisfied, the fact that ϕ-ing is a way to Z causes, via a causal process that involves

---

[23] It may look here like my argument presupposes that counterfactuals are transitive. But it can be re-expressed without loss (though at the cost of wordiness) to use the valid argument form "A → B; (A and B) → C; therefore, A → C" (where "→" is the counterfactual conditional connective).

natural selection, X to ϕ. I do not see what is so special about natural selection. Why can't there be other kinds of causal processes leading from the fact that ϕ-ing is a way to Z, to X's ϕ-ing, that are sufficient for it to be the case that X ϕ-ed in order to Z? The right theory, I believe, will say something more general about what kind of causal connection must exist between the fact that ϕ-ing is a way to Z, and X's ϕ-ing on this occasion. (Maybe selectional processes are the only kind that satisfy this as-yet-unformulated general description in the actual world.)

Here is a stab at a theory like that, suggested to me by Kieran Setiya. Maybe we could say

TELEOLOGY-4: The reason why X ϕ-ed was to Z if and only if
  (i) one cause of X's ϕ-ing was that X was disposed to ϕ in that kind of situation (and, in ϕ-ing, X manifested this disposition);
  (ii) one cause of X's having this disposition is that X was of kind K, where Ks (generically) are disposed to ϕ in that kind of situation;
  (iii) one cause of the fact that Ks are disposed to ϕ in that kind of situation is that ϕ-ing in that kind of situation is a way for Ks to Z.

Clause (iii) is where the fact that ϕ-ing is a way to Z is given its causal role. Here we have a unified, non-disjunctive theory. Does it cover both plants closing their stomata and Aristotelian rocks moving down? Let's check. A given plant closed its stomata; in doing this it manifested a disposition to close its stomata at night; it has this disposition because it is a (species of) plant; and one cause of the fact that plants (of that species) are disposed to close their stomata at night is that closing their stomata at night is a way for plants of that species to conserve water. That all sounds right. Now rocks: a given Aristotelian rock moved down; in doing this it manifested a disposition to move down, when downward is the direction of the center of the universe; it has this disposition because it's made of earth; and one cause of the fact that things made of earth have this disposition is the fact that moving down when downward is the direction of the center of the universe is a way for things made of earth to get to the center of the universe.

Everything was going well until that last sentence. But that last sentence doesn't seem right. The fact that things made of earth are disposed to move down, when downward is the direction of the center of the universe: this fact doesn't seem to me to have *any* causes. Certainly the fact that moving down is a way to get to the center doesn't seem to be a cause. We didn't run into this kind of trouble with plants, since the dispositions a given species has can change over time, and can be *caused* to change.

I seem to have shot myself in the foot. TELEOLOGY-1 did really well with the rocks example. But in revising it to arrive at TELEOLOGY-4 the causal role that the fact that ϕ-ing was a way to Z needed to play, for some behavior to be directed at Z-ing, has changed. It has changed so much, in fact, that now, even though the fact that moving down is a way to get the center caused the rock to move down, its causing the rock to move down is not enough to make it true that the rock moved down in order to get to the center.

All hope, however, is not lost. There is a way to fix clause (iii) so that it does cover rocks and plants. Make it more liberal, so that it requires only that the fact that ϕ-ing is a way to Z be a cause *or a ground* of the fact that Ks are disposed to ϕ. For while I don't think that the fact that moving down is a way to get to the center is a cause of the fact that things made of earth are disposed to move down, when that's the direction to the center, I do think it's a ground. By letting in grounds as well as causes, the more liberal version of clause (iii) can be stated in terms of reasons why. Here's what it looks like:

(iii*)   one *reason why* Ks are disposed to ϕ in that kind of situation is that ϕ-ing in that kind of situation is a way for Ks to Z.

Let's now understand TELEOLOGY-4 to include this amended clause.

So amended, TELEOLOGY-4 is not exactly a causal theory of non-finite reasons. But it does still achieve the goal of reducing non-finite reasons why to factual reasons why.

The theory TELEOLOGY-4 does not mention natural selection; but natural selection is still relevant to the plant's meeting the conditions in TELEOLOGY-4: natural selection appears in the answer to the question *why clause (iii) is satisfied in this case*—the question why it is that (plants are disposed to close their stomata at night because closing their stomata at night is a way for them to conserve water). The answer to *this* why-question is that there was selection for water conservation in the ancestral environment.

I want to leave the topic of teleological answers to why-questions that do not give an agent's reasons for acting. First, though, I should acknowledge that I haven't yet said anything about functions, even though most of the literature by philosophers on teleology in biology aims to analyze the notion of a function, not to give a theory of teleological explanations. I don't think one needs a theory of functions in order to have an adequate theory of teleological explanation. True, there is some connection between the fact that the function of my heart is to circulate my blood, and the fact that my heart is pumping in order to circulate my blood. But I don't think functions need to be mentioned explicitly in a theory of reasons why that accounts for the fact that the reason why my heart is pumping is to circulate my blood. In fact, it is tempting to say that the notion of a function is to be analyzed, at least in part, in terms of non-finite reasons why: a function of this X is to Z only if there is some ϕ such that the reason why Xs (generically) do ϕ is to Z.[24]

6.2 Reasons for Action

When someone does something intentionally, we can ask why they did it, and mean this question as a request for their reasons for doing that thing.

This is not the only meaning "Why did he ϕ?" can have; even when someone acts intentionally, we can ask why he did that thing and mean our question as a request for causes of his action that need not have anything to do with his reasons for acting. I might shout intentionally, and my reason for shouting might be that I have been insulted. Still, someone might ask a third party why I shouted and accept "because he is overtired" as a good answer.

---

[24] In this schema "X" holds the place for a common noun, like "heart," or "kidney," or "paperweight." A slightly different schema is needed to cover attributions of functions to kinds, as in "The function of *the heart* is to circulate the blood."

Some philosophers attribute functions to kinds of behaviors, like "eggshell-removal behavior"; unlike hearts or kidneys, behaviors are kinds of events. Allen and Bekoff for example report that "a function of eggshell removal behavior in birds is to protect their offspring" ("Biological Function, Adaptation, and Natural Design," p. 611). If one wants to take attributions of functions to behaviors seriously, one needs a separate schema for them: something like, the function of behavior B (in things of kind K) is to Z only if Ks engage in B in order to Z.

Wright defended a simpler theory of functions, roughly equivalent, given the rest of his theory, to "the function of this X is to Z iff this X is there in order to Z" (*Teleological Explanations*, p. 81). Boorse, in "Wright on Functions," also analyzed functions in terms of goals or ends.

There are a variety of examples that stand in the way of making the conditional in the text into a biconditional. Suppose that a rock moves down in order to get to the center of the universe; is it a function of the rock to get to the center of the universe?

The topic of this section is answers to why-questions that give an agent's reasons for acting. I am including a discussion of this topic for two reasons. First, for the sake of completeness; with it, I will have said something about all kinds of answers to why-questions about events (I take intentional actions to be kinds of events). More importantly, though, I am including it in order to illustrate the importance of the distinction between higher- and lower-level reasons outside of the place where I introduced it. This distinction has an important role to play in a theory of acting for reasons.

The notion of an agent's reason for acting is not the same as the notion of a reason for her to act. That P could have been my reason for ϕ-ing even if there were no reasons for me to ϕ. That it would quench my thirst could have been my reason for drinking the contents of the glass even if there had been no reason for me to drink (say, because the glass contained gasoline, which would not quench my thirst). Similarly, it can happen that there are several reasons for me to ϕ, and I do ϕ, but none of them is my reason for ϕ-ing.

One difference between these two kinds of reasons is this: if some consideration was X's reason for ϕ-ing, then it follows that X ϕ-ed; not so if some consideration was a reason for X to ϕ.[25]

I am here interested in a person's reasons for acting—the reasons for which he acted—not the reasons there were for him to act. As I said, there is a connection between the question why X ϕ-ed and X's reasons for ϕ-ing—the question can be used to request those reasons. Is there some connection between some why-question and the reasons there are for X to ϕ? Maybe, but if so the nature of the connection is controversial. The least controversial claim one could make is this: if R is a reason for X to ϕ, and moreover X *should* ϕ, then R is part of the answer to the question why X should ϕ. This why-question asks, not why some event occurred or why X in fact ϕ-ed but why some normative fact obtains.[26]

---

[25] Setiya first drew my attention to this difference between the two kinds of reasons, in his paper "Reasons and Causes." Some authors distinguish between explanatory reasons and justifying reasons. Others distinguish between motivating reasons and normative reasons. There is some controversy about whether these are the same distinction, and are the same as the distinction between X's reasons for ϕ-ing and the reasons for X to ϕ; for some discussion see Lenman, "Reasons for Action: Justification vs. Explanation." To avoid this controversy I will not use the "explanatory/justifying" or the "motivating/normative" terminology.

[26] Certainly there can be reasons for X to ϕ even if X should not ϕ (those reasons are outweighed); so it is false in general that if R is a reason for X to ϕ, then R is part of the

I will be looking at the role the distinction between levels of reasons should play in accounts of what it takes for a consideration to be the reason for which someone acted. Before doing that I should remark that the distinction also appears in accounts of what it takes for a fact to be a reason for someone to act. Mark Schroeder defends the view that R is a reason for X to φ iff there is some proposition p such that (i) X wants it to be the case that p is true, and (ii) R is a reason why X's φ-ing "promotes" the truth of p, where "promotes the truth of p" means something like "would help bring about the truth of p" (*Slaves of the Passions*, chapter 2). So, on this view, the fact that the store sells milk is a reason for Jones to go to the store, since (i) Jones wants it to be the case that he has some milk, and (ii) the fact that the store sells milk is a reason why going to the store would help bring it about that he has some. This theory is a version of the idea that all reasons to act "have their source in" the agent's desires. Schroeder cites several philosophers who object that if the fact that R is a reason for X to φ has its source in one of X's desires, then the fact that X has that desire must *also* be a reason for X to φ—which, they claim, need not always be so. Schroeder replies, in effect, that these philosophers are assuming that higher-level reasons are also lower-level reasons: assuming that if A is a reason why ⟨R is a reason for X to φ⟩, then A a reason for X to φ.[27] Schroeder claims that this assumption is false. Suppose that Ronnie

---

answer to the question why X should φ. (Toulmin, in *Reason in Ethics*, held that something is a reason for X to φ iff it is a reason why X ought to φ; Schroeder objects on the grounds I just gave in *Slaves of the Passions*, p. 35.)

What about the other direction? If X should φ, so the question of why he should φ arises, is every part of the answer to the question a reason for X to φ? This has been disputed. Schroeder disputes it in *Slaves of the Passions,* as does Setiya, in "What is a Reason to Act?": that X wants to Z can be part of the answer to the question why he should φ, without being a reason for him to φ. (It is maybe worth noting that John Broome *defines* "reason for X to φ" as a certain kind of part of the answer to the question why X should φ in his book *Rationality Through Reasoning*—though this is compatible with the falsity of the above conditional.)

[27] This is not quite the thesis I rejected when I asserted (High ↛ Low) in chapter 4; in that principle, the first-level reasons were reasons why something happened, not reasons for an agent to do something.

Schroeder sometimes has his opponents assume that if A is a reason why ⟨R is a reason for X to φ⟩, then A is part of R ("there is only one way for the fact that Ronnie desires to dance to explain why there is a reason for him to go to the party. It is for this fact to be part of the reason for Ronnie to go to the party" (p. 23)). Schroeder probably has his opponents assume this because some of them have actually put it this way. He heaps ridicule on this assumption: "it denies a distinction with respect to reasons that is a perfectly good distinction in many other domains. Although being inaugurated is a necessary condition for someone to be president of the United States, and the fact that George W. Bush was inaugurated less than

wants to dance. That he wants to dance is reason why ⟨the fact that there will be dancing at the party is a reason for Ronnie to go to the party⟩; but, Schroeder claims, that Ronnie wants to dance is not a reason for him to go to the party.[28]

Let's get back to theories of acting for reasons. Just as, when a plant closes its stomata, there may be both factive and non-factive reasons why it closed its stomata, when someone acts for a reason, there may be both factive and non-factive reasons for which he acts. Jones goes to the store; one of his reasons for going was that the store sold milk (this is the factive reason), and another of his reasons was to buy some milk (this is the non-factive one). Non-factive reasons give the agent's purpose or aim in acting.

At minimum a theory of acting for reasons should fill in the right-hand sides of these schemas:

(2)   One of X's reasons for ϕ-ing was that P iff. . .

(3)   One of X's reasons for ϕ-ing was to Z iff. . .

I say "at minimum" because a theory of acting for reasons should also cover other locutions for giving an agent's reasons for action, for example, "X ϕ-ed on the ground that P," and "X ϕ-ed in order to Z." But I will focus just on (2) and (3).[29]

There is a tradition in the philosophy of action of asserting a close connection between someone's reasons for acting and the causes of their action;[30] I don't expect you will be shocked when I say that I adhere to that tradition. My aim here is not to defend a novel causal theory of acting for reasons but to show that the most promising causal theories of reasons for action make use of the distinction between levels of reasons why.

four years ago is part of why he is president, the fact that Bush was inaugurated isn't part of the president of the United States" (p. 24). The assumption I put in the main text cannot be ridiculed in this way; maybe it is the thesis Schroeder's opponents should defend.

[28] Jonathan Dancy also invokes something like a distinction between levels of reasons in his theory of reasons to ϕ. Reasons to ϕ, he holds, are facts that favor ϕ-ing; they must be distinguished from "enablers," which (merely) enable a fact to favor ϕ-ing without themselves favoring ϕ-ing. (See *Ethics Without Principles*, chapter 3; I compared the notion of an enabler to the notion of a higher-level reason why in footnote 19 in chapter 4.)

[29] "X ϕ-ed in order to Z" appears to be equivalent to "X's reason for ϕ-ing was to Z"—though see footnote 37. I will say something about one other way to give an agent's reasons in the appendix to this chapter.

[30] One important member of this tradition is Davidson; see for example "Actions, Reasons, and Causes." I will discuss some of Davidson's ideas below.

A causal theory of acting for reasons claims that whether an agent acts for a reason is a matter of how the act is caused. The obvious suggestion is that acting for a reason requires the act to have certain mental causes, and one direction of a connection between the act and its mental causes is fairly uncontroversial:

(4) One of X's reasons for ϕ-ing was that P only if the fact that X believed that P was a cause of X's ϕ-ing.

(5) One of X's reasons for ϕ-ing was to Z only if the fact that X wanted to Z was a cause of X's ϕ-ing.

Can the right-hand sides of (4) and (5) be turned into conditions that are sufficient as well as necessary? The simplest idea is just to replace "only if" with "iff" in (4) and (5):

THE SIMPLEST THEORY OF ACTING FOR REASONS:
One of X's reasons for ϕ-ing was that P iff the fact that X believed that P was a cause of X's ϕ-ing.

One of X's reasons for ϕ-ing was to Z only iff the fact that X wanted to Z was a cause of X's ϕ-ing.

If The Simplest Theory is right, then answers to why-questions that request reasons for action do almost the same thing as answers to why-questions that do not: both provide causes.[31] It is just that in the case of an agent's reasons for action, the agent's reasons (that P, for example) are not themselves causes; rather the cause is one of the agent's mental states, one that has the same content as the reason. If his reason is that P, the cause is his belief that P; if his reason is to Z, the cause is his desire to Z.

Unfortunately The Simplest Theory is false; the conditions on the right-hand sides are not sufficient. There are many proposals for salvaging them, based on different diagnoses of their flaws. I am going to look at two proposals, and show that my distinction between levels of reasons has an important role to play in defending each of them.

Thomas Nagel imagined someone who has been conditioned to put a coin in a pencil sharpener whenever he believes that there is a pencil sharpener within reach and wants to have something to drink.[32] The conditioning ensures that his belief and desire cause his action. But, it

---

[31] Or grounds, of course; but I am mostly ignoring reasons that are grounds here.
[32] *The Possibility of Altruism*, pp. 33–4.

seems, it is false that one of his reasons for putting the coin in the pencil sharpener was to get a drink. And it seems false that one of his reasons for putting the coin in the pencil sharpener was that there was a pencil sharpener in reach. So this is a counterexample to The Simplest Theory.

Nomy Arpaly and Timothy Schroeder claim that The Simplest Theory can be repaired by requiring, not just that the belief, or the desire, cause the action, but also that (i) the belief, or the desire, play a role in "rationalizing" the action, and that (ii) the action be caused in part by the fact that the agent has this rationalization for it. That there is a pencil sharpener in reach, and that one wants something to drink, do not rationalize putting a coin in the pencil sharpener. In my terms, a preliminary formulation of their theory looks like this (the reason why it is a preliminary formulation, and is not their final view, will become clear presently):[33]

RATIONALIZATIONS (preliminary):

(i) One of X's reasons for ϕ-ing was that P iff the fact that X believed that P was a cause of X's ϕ-ing, *and* this belief, possibly along with other mental states[34] (a desire, for example), rationalized ϕ-ing, *and* the fact that X had this rationalization for ϕ-ing was a cause of X's ϕ-ing.

(ii) One of X's reasons for ϕ-ing was to Z iff the fact that X wanted to Z was a cause of X's ϕ-ing, and this desire, possibly along with other mental states (a belief, for example), rationalized ϕ-ing,[35] and the fact that X had this rationalization for ϕ-ing was a cause of X's ϕ-ing.

I won't try to say very much about what it takes for a collection of mental states to rationalize an action. Certainly it is sufficient for them to

---

[33] Arpaly and Schroeder present their theory in *In Praise of Desire*, pp. 62 ff. (they state their theory as a theory of what it is to act for a reason, not, as I will, as a theory of what it is for it to be the case that one's reason is that P, or to Z). They cite Arpaly, *Merit, Meaning, and Human Bondage*, and Wedgwood, "The Normative Force of Reasoning," as the first to propose this view, and discuss it further in "A Causal Theory of Acting for Reasons." Much earlier Davidson required the contents of one's belief and/or desire to rationalize the action, but did not require the fact that they did this to also be a cause of the action ("Actions, Reasons, and Causes").

[34] Some philosophers hold that a belief, say the belief that one ought to ϕ, can rationalize ϕ-ing all by itself; (i) is worded so as to be compatible with this view.

[35] In light of footnote 34, I should probably also allow the case where the fact that X believed that he ought to Z, and the fact that X believed that ϕ-ing was a way to Z, were causes of X's ϕ-ing. Since this qualification is not relevant to what I want to say about RATIONALIZATIONS, I will ignore it from now on.

rationalize an action that it is rational for someone who has them to perform that action; but this is not necessary.[36] A vague, intuitive grip on rationalization is enough for what I want to say.

Presumably if the desire clause (ii) speaks of combines with a belief B to rationalize ϕ-ing, then B will either be, or entail with the agent's other beliefs, that ϕ-ing is a way for the agent to Z; at least this is often the case. It is here that this theory of when some end is someone's reason for acting connects up with my theory(/ies), from the previous section, of non-finite reasons why that are not reasons for action. The most direct connection is to the theories TELEOLOGY-1 through TELEOLOGY-3. About the cases they were meant to cover, cases where the end of a piece of behavior is not a reason for which the agent performed the behavior, those theories say that one of the causes of the performance was the fact that performing that behavior was a way to achieve the end; here, where the end *is* a reason for which the agent performed the behavior, the *fact* that performing that behavior was a way to achieve the end need not be a cause (it may not even be a fact; the behavior may be completely misguided, not a way to achieve the end even under idealized conditions), but the agent's *belief* that performing that behavior is a way to achieve the end will (in most cases at least) be a cause.[37]

RATIONALIZATIONS, however, is still not right; it falls prey to the problem of "deviant causal chains." Donald Davidson illustrated this problem with his famous mountain climber case:

> Beliefs and desires that would rationalize an action if they caused it in the *right* way—through a course of practical reasoning, as we might try saying—may cause it in other ways. If so, the action was not performed with the intention that we could have read off from the attitudes that caused it. What I despair of spelling out is the

---

[36] For more on rationalization, see Wedgwood, "The Normative Force of Reasoning," p. 662, or Arpaly and Schroeder, *In Praise of Desire*, pp. 62 ff.

[37] As I said, you can give an agent's reason for acting by using "in order to": often, "Jones went to the store in order to buy some milk" is equivalent to "One of Jones's reasons for going to the store was to buy some milk." But we should leave open the possibility that a piece of intentional behavior has ends other than the agent's. A Marxist might want to say that I voted for the socialist candidate in order to hasten the coming of the revolution, even if I myself have no desire to overthrow the government, even if my reason for doing it was just to get a raise. (I suspect that "in order to" is actually ambiguous between a meaning that gives an agent's reason, and a meaning that doesn't; I defend the corresponding ambiguity claim about "reason why" in Appendix D.)

way in which attitudes must cause actions if they are to rationalize the action.

Let a single example serve. A climber might want to rid himself of the weight and danger of holding another man on a rope, and he might know that by loosening his hold on the rope he could rid himself of the weight and danger. This belief and want might so unnerve him as to cause him to loosen his hold, and yet it might be the case that he never *chose* to loosen his hold, nor did he do it intentionally.

("Freedom to Act," p. 79)

If we understand this case so that what unnerves the climber is that he has a belief–desire pair that rationalizes dropping the rope, then it is a counterexample to RATIONALIZATIONS.

The justification for calling this problem the problem of "deviant causal chains" is the idea that the chain of causes leading from the climber's possession of a rationalization for dropping the rope to his dropping of the rope is the "wrong kind" for that rationalization to contain his reasons for dropping the rope. With this terminology in place, the question becomes, what distinguishes the deviant causal chains from non-deviant ones?[38] Many philosophers have taken a stab at solving the problem of deviant causal chains. I think the solution will make use of the distinction between higher- and lower-level reasons.

Here is how Arpaly and Schroeder articulate the strategy they favor for fixing RATIONALIZATIONS:[39]

... one wants to distinguish the case in which a flying brick breaks a window in virtue of the brick's mass, velocity, and so on from the case in which a demon observes the mass, velocity, and so on of the brick and then breaks the window itself.... one aims to explain why the former case is one in which the flying brick genuinely breaks the window in virtue of its mass, velocity, and so on, whereas the

---

[38] I will shortly suggest that this standard name for the problem points us away from the solution.

[39] Something like this idea also appears in, for example, Schlosser, "Basic Deviance Reconsidered," and Wedgwood, "The Normative Force of Reasoning." Technically, what follows is offered as a solution to the problem of deviant causal chains in "basic action," actions not done by doing anything else. The problem of deviant causal chains in non-basic action is easier to solve; see Setiya, *Reasons Without Rationalism*, p. 32, for one solution.

latter case is one in which the flying brick does *not* genuinely break the window in virtue of its mass, velocity, and so on. Solving this metaphysical problem is a perennial challenge. But it is not specific to action theory.

We think it can be taken for granted that there is a principled difference between bricks breaking windows in virtue of their masses and demonically mediated window breaking. If there is, then this principled difference can be imported to explain the difference between the rationalizing attitudes in Davidson's mountain climber causing his release of the rope in virtue of their rationalizing a release and the same attitudes causing his release of the rope via some other causal pathway that is not truly (not merely?) in virtue of the existence of the rationalization. With this difference marked, Davidson's mountain climber becomes . . . someone with a rationale for a behavior who performs the behavior in a way ultimately caused by the rationale, but not caused in virtue of the properties of the rationalizing attitudes that makes them into rationalizers.

(pp. 70–1)

Just what is the distinction Arpaly and Schroeder are drawing? Certainly there is a distinction between a situation in which a brick breaks a window, and one in which a demon observes the state of the brick, and then breaks the window himself. The difference is in which thing breaks the window. This distinction can't be the one they want to draw, though. True, in the climber example, the climber's rationalization doesn't drop the rope. (It's not even the sort of thing that *could* drop a rope.) But in the "good case," where the climber does not become nervous, but drops the rope intentionally,[40] it is still false that the rationalization drops the rope. In both cases the climber drops the rope.

Although speaking of "bricks breaking windows" suggests a focus on what (thing) breaks the window, it is clear that this is not really what Arpaly and Schroeder have in mind. Their use of "in virtue of" suggests a Davidsonian way of thinking about causation, on which causation is primarily a relation between events, and whenever an event E1 causes an event E2, we can ask which of E1's properties are the properties in virtue of which E1 caused E2. But here I don't see a useful distinction either. Here's

---

[40] This case is not "good" in the moral sense, of course.

an event E that caused the breaking in the "good" (non-demonic) case: the event consisting of a brick with mass M flying toward the window. This event is also a cause of the breaking in the bad (demonic) case. The passage from Arpaly and Schroeder suggests the thesis that in the good case but not the bad case, E causes the breaking in virtue of being a flying of a massive brick. This thesis looks false. In *both* cases, E causes the breaking in virtue of being a flying of a massive brick. After all, in both cases the counterfactuals are the same: if E had not had the property of being a flying of a massive brick, the window would not have broken. (We can assume that in the bad case the demon only responds to massive bricks.)

The same, I think, goes for the climber. Let F be the event consisting in the climber's possessing a rationalization for dropping the rope. In both the good and the bad case, the property of F in virtue of which it causes the climber to drop the rope is the property of being an event consisting in the climber's possessing a rationalization.

Although Arpaly and Schroeder's claims about the properties in virtue of which the brick breaks the window do not seem to me to solve the problem, their use of "in virtue of" does suggest another approach. Let us switch from speaking of events as causes to speaking of facts as causes (as I prefer). Although this is not, I think, how they understand the question, we can hear the question of which properties of the cause are those in virtue of which it caused the effect as a request for the reasons why the cause is a cause—as a request for higher-level reasons. This, I think, is a much better question to ask.

I want to suggest that the difference between acting on a rationalization for that action, and (merely) being caused to act by that rationalization, can be spelled out in terms of the (higher-level) reasons why the rationalization caused the act.

If I am right, then saying that the bad cases involve "deviant causal chains" is misleading, because it suggests that the difference between the good and bad cases can be discerned by looking at the chain of causes leading from the rationalization to the action. If the difference involves higher-level reasons, then in principle there could be two indiscernible causal chains, both terminating in someone's doing something, only one of which terminates in someone's acting for a reason—as long as the chains differed over the reasons why some of the causes in the chain were causes.

So why, in the bad, deviant, case, is the fact that the climber had a rationalization for dropping the rope a cause of the fact that he dropped the rope? The most salient answer is: because his having the rationalization caused him to become nervous. In the good case, on the other hand, this is not the answer.

But this does not help very much. I think there could be a climber for whom nervousness was a essential part of the process of his acting for reasons. For a climber like that, the fact that he became nervous would be a reason why his having a rationalization caused him to drop the rope, in both the good and the bad case.[41]

The fact that the climber became nervous is a link in the causal chain from the fact that he has a rationalization for dropping the rope, to his dropping of the rope. The fact that he became nervous is also a higher-level reason why: it is a reason why the fact that he had a rationalization caused him to drop the rope. My hunch is that the difference between the good case and the bad case lies in higher-level reasons why that lie outside the causal chain from the rationalization to the act; so, if my hunch is right, his becoming nervous is the wrong kind of higher-level reason to look at.

What other high-level reasons are there in this case? That's a good question, but I think a better question is what other high-level reasons there *aren't*. It is tempting to say, about the bad case, not only that the fact that it made him nervous is a reason why the rationalization caused the climber to drop the rope, but also that this is the *only* (relevant) reason why it caused him to drop the rope. Now, I think, we are getting somewhere. For even if we imagine a version of the good case inhabited by a climber for whom becoming nervous is an integral part of acting for reasons, it does not seem right to say that his becoming nervous is the only relevant reason why his having a rationalization caused him to drop the rope. One also wants to say, in this version of the good case, that the fact that he has a general disposition to ϕ when he has a rationalization for ϕ-ing is also a reason why he dropped the rope.

---

[41] I am thus skeptical of attempts to solve the problem of deviant causal chains by saying that a chain is non-deviant only if the fact that the agent has a rationalization is a "direct cause," in some sense, of the agent's act. (Arpaly and Schroeder advocate a solution like this in section 4 of "A Causal Theory of Acting for Reasons," though they revise the solution later in the paper.)

Davidson's climber also has this disposition, of course; but in the original version of the bad case, the fact that the climber had this disposition is not a reason why his having a rationalization caused him to drop the rope.

We are close, but not quite there. An extremely sophisticated climber could notice, not just that he has a rationalization for dropping the rope but also that he is in general disposed to ϕ when he has a rationalization for ϕ-ing, and it could be his noticing both of these things that makes him nervous.

There is still a difference, though, between this version of the bad case and the good case, in which the climber drops the rope intentionally. In the good case, the reason why the climber's having a rationalization causes him to drop the rope is that it causes him to *manifest* his disposition. That doesn't happen in the bad case, nor do I see how to construct a bad case in which it does.

The theory of acting for reasons we end up with is not that new,[42] but the path to it, through higher-level reasons why, does, I hope, cast it in a different light:

RATIONALIZATIONS (final):

(i) One of X's reasons for ϕing was that P iff the fact that X believed that P was a cause of X's ϕ-ing, and this belief, possibly along with other mental states (a desire, for example), rationalized ϕ-ing, and the fact that X had this rationalization for ϕ-ing was a cause of X's ϕ-ing, *and the reason why this rationalization caused X to ϕ is that it caused X to manifest this disposition: the disposition to do those things for which one has a rationalization.*

(ii) One of X's reasons for ϕing was to Z only iff the fact that X wanted to Z was a cause of X's ϕ-ing, and this desire, possibly along with other mental states (a belief, for example), rationalized ϕ-ing, and the fact that X had this rationalization for ϕ-ing was a cause of X's ϕ-ing, *and* the reason why this rationalization caused X to ϕ is that

---

[42] It is closely related to Christopher Peacocke's solution to the problem of deviant causal chains in chapter II of *Holistic Explanation*, and to Ralph Wedgwood's theory of believing for reasons in "The Normative Force of Reasoning;" Arpaly and Schroeder may consider it a rephrasing of the final version of their theory, implicit in the quotation from them above, according to which one acts for a reason iff one's action is caused by one's beliefs and/or desires in virtue of the fact that those beliefs and/or desires rationalize that action (*In Praise of Desire*, p. 62).

it caused X to manifest this disposition: the disposition to do those things for which one has a rationalization.

To come full circle, Arpaly and Schroeder take the example of a brick causing a demon to break the window to be relevantly like an example of a deviant causal chain. From our perspective here, it does not look relevantly similar. For while windows have dispositions to break when struck, they do not (also) have (separate) dispositions to break when struck by bricks. So when the brick strikes the window it doesn't cause the window to manifest a disposition that is not manifested when the demon breaks it.

The theory of acting for reasons I have been examining is open to objections. Can't someone act for a reason, without having beliefs and desires that in any sense rationalize the action? Can't my reason for poking my brother be that he is smiling, even if I don't have any mental states that combine with my belief that he is smiling to rationalize poking him? Kieran Setiya thinks the answer is yes (to the general question at least, if not to the question about this particular example), and in part 1 of *Reasons Without Rationalism* proposes a causal theory of acting for reasons that does not require rationalizations. Like the theory I just discussed, though, in Setiya's theory whether someone acts for a reason depends on facts about higher-level reasons: it depends on whether the reasons why his beliefs and/or desires cause his action are the right sorts of reasons.

Arpaly and Schroeder looked at The Simple Theory, the "upgrades" of (4) and (5) to biconditionals:

One of X's reasons for ϕ-ing was that P iff the fact that X believed that P was a cause of X's ϕ-ing.

One of X's reasons for ϕ-ing was to Z only iff the fact that X wanted to Z was a cause of X's ϕ-ing.

and decided that they couldn't be right because they allow someone to do something for a reason even in cases where that person's beliefs and desires do not rationalize the action. Setiya sees different flaws: according to The Simple Theory, someone can do something for a reason even in cases where that person does not know that he is doing that thing; what's more, it can be that one of X's reasons for ϕ-ing was that P even when X did not know that he was doing it for that reason; and it can be that one of X's reasons for ϕ-ing was to Z even when X did not know that he was doing it for that reason. Nagel's pencil sharpener case illustrates the latter two flaws. Since the man in the example was conditioned to insert coins upon having the right beliefs and desires, it seems perfectly possible that if we were to ask him why he is putting a coin into the pencil sharpener, he would reply that he has no idea why he's doing it. Anyway, there is certainly no guarantee, from the description of the case, that he will say, or believe, that he is doing it in order to get a drink. But if he does not even believe that he is doing it in order to get a drink, how can it be that that is one of his reasons for doing it?

Setiya holds—and I find this claim plausible—that, necessarily, whenever one does something intentionally, he knows that he is doing it; and that necessarily, whenever someone does something for a reason, he knows what his reason is.[43] What needs to be added to the right-hand sides of (4) and (5) so that these necessary truths follow from them?

It will do for us to just look at Setiya's analysis of "One of X's reasons for ϕ-ing was that P." Here is my statement of his theory:[44]

- One of X's reasons for ϕ-ing was that P iff (i) X believed that P, (ii) X had a desire-like belief that ⟨he was ϕ-ing because he believed that P⟩, and (iii) this belief, and this desire-like belief, were (non-deviant) causes of X's ϕ-ing, and (iv) *the desire-like belief did not just cause X's ϕ-ing but also caused X's belief that P to cause X's ϕ-ing.*[45]

The desire-like belief mentioned in (ii) is, Setiya claims, an intention. On his theory, in order to act for the reason that P, one must have a desire-like belief the content of which is that one is ϕ-ing because one believes that P. Desire-like beliefs are beliefs; and one cannot believe that ⟨one is ϕ-

[43] Most of Setiya's focus is on the first idea, which he gets from Anscombe (*Intention*, p. 11). My statement of the claim is slightly oversimplified, but the complications do not matter here. Setiya discusses them in *Reasons Without Rationalism*, p. 25, and "Practical Knowledge."

[44] See *Reasons Without Rationalism*, pp. 43–8.

[45] Clause (ii) is not exactly as Setiya put it. The true version is more complicated, and my focus is on clause (iv) not clause (ii), but for the record Setiya actually requires the desire-like belief to have this content (what follows is my own wording): he was ϕ-ing because he believed that P, and the reason why he was ϕ-ing because he believed that P was that he had this very desire-like belief (see *Reasons Without Rationalism*, pp. 43 ff).

In clause (iii) Setiya's theory presupposes a solution to the problem of deviant causal chains; his theory as a whole requires a solution different from the one discussed above, that appeals to rationalizations. He sketches a solution on p. 32, but the details of that solution are not important here.

ing because one believes that P) without also believing that one is ϕ-ing. That is (part of) how Setiya's theory secures the first necessary truth, that one cannot do something intentionally without knowing that one is doing that thing.[46]

It is, however, (iv) that I really care about.[47] For it is in condition (iv) that Setiya's theory implicitly appeals to the distinction between higher- and lower-level reasons.

Condition (iv) speaks of an intention's causing a belief to cause an action. But this way of putting it does not do a very good job capturing what I think Setiya had in mind. If the cue ball hits the one ball, which hits the two ball, which goes into the pocket, then the cue ball caused the one ball to cause the two ball to go into the pocket. Here the "cause of the cause" is an earlier event in the causal chain. But the role for the intention that condition (iv) is supposed to articulate is not like this; having the intention is not supposed to cause the agent to believe that P. The agent probably already believes that P. Instead, (iv) is supposed to assign the intention the role of "causing" the (already existing) belief that P to *have the power to cause the agent to act*. The intention is supposed to "raise the status" of the belief from that of something that merely preceded the action, to a cause of the action. But "cause" is not the best word for these "status-raisers." The better term, I think, is "reason why."[48] So I think that condition (iv) is most perspicuously rendered as

- The fact that X had the desire-like belief in (ii) is *the reason why* X's belief that P was a cause of X's ϕ-ing.

Clause (iii) requires the belief and the intention to each be a cause of the agent's act. So why does the intention *also* need to be a reason why the belief is a cause? It needs to do this to guarantee that the agent always knows the reasons on which he is acting. Since this is not a treatise on acting for reasons, I will only give part of the argument for this claim.

---

[46] So far we have just seen why one must *believe* that one is ϕ-ing, when one is ϕ-ing for a reason. The question of why this belief constitutes knowledge cannot be answered just by looking at the theory as I have written it; for discussion see Setiya's "Practical Knowledge."

[47] Setiya's statement of (iv) appears in note 36 on page 43 of *Reasons Without Rationalism* (see also page 59); my statement of it is worded differently, but is essentially the same.

[48] The term "status-raiser" comes from Yablo, "Advertisement for a Sketch," p. 99; he also uses "ennobler" for status-raisers. I claimed in chapter 4 that if some fact ennobles an event into a cause of C, then that fact is a reason why C happened.

Speaking abstractly: if I am ϕ-ing, and I intend to be $\langle\phi$-ing because I believe that P$\rangle$, then the belief that constitutes this intention constitutes knowledge only if it is "no accident" that it is true. So it has to be no accident that I truly believe that $\langle$I am ϕ-ing because I believe that P$\rangle$. One way for this to be no accident is for the intention to "help make it the case" that its content is true: for it to help make it the case that my belief that P is a cause of my ϕ-ing. And to "help to make it the case that" my belief is a cause of my ϕ-ing is just to be a reason why my belief is a cause of my ϕ-ing.[49]

Appendix D: "Reason why …" is Ambiguous

There is one loose end I want to tie up. I have focused on reports of an agent's reasons for acting of the form "X's reason for ϕ-ing is …" But, as I said, there are many ways to report an agent's reasons for acting. One way uses the form of words I spent chapters 2 through 5 theorizing about: "reason why." If the store really does sell milk, then "The reason why John went to the store was that it sold milk" can be heard as equivalent to "John's reason for going to the store was that it sold milk."[50] This raises the question of whether what I said in the last section is consistent with the theory of reasons why defended in the rest of this book. For while I did not endorse either of the theories of acting for reasons I discussed, I did not rule them out as inconsistent with my other committments either.

And there is an argument to be made that both are in fact inconsistent with (T1f). Make things as promising as possible for the hope that they are consistent with (T1f): suppose that the fact that the store sold milk was a cause of John's going to the store, because it caused him to believe that the store sold milk, and this belief was among the causes of his going to the store. Nevertheless, (T1f) says that all it took for "The reason why John went to the store was that it sold milk" to be true was for this fact to

---

[49] A more complete argument would have to establish that no other way of ensuring that it is no accident that the content of the intention is true will work.

[50] Without the background assumption that the store sold milk, they may not be equivalent: in such a context "John's reason for going was that it sold milk" can still be true, but it is hard for me to hear "The reason why John went to the store was that it sold milk" as having a true reading. I am not sure whether this is for semantic or for pragmatic reasons, though.

be a cause; while both of the theories of acting for reasons I discussed say that more was required.

My response is that "reason," when it occurs in contexts like "reason why John went to the store," is ambiguous. For certainly the sentence

(6)   The reason why John went to the store was that it sold milk.

has two readings. Read one way, it reports John's reason for going. Read another way, it could be true even if John did not go to the store for any reason, did not go intentionally. On this second reading, (6) could be true if the fact that the store sold milk caused a chip in John's brain to cause him to go to the store "against his will." I hold that these two readings are due to an ambiguity in "reason." Theory (T1f) is only meant to be true on one disambiguation, the meaning "reason" has when one makes no claim to be reporting an agent's reason for acting. The theories in section 5.2 are meant to be true on the other disambiguation. I hold, for similar reasons, that "in order to" and "because" are ambiguous between meanings used to report someone's reason for acting, and meanings used when one makes no claim to be reporting someone's reason for acting.

I need to shore up my argument that "reason" is ambiguous. It is not enough to observe that (6) has two readings; there are other possible sources of multiple readings besides an ambiguity in "reason." The most salient alternative is contextual restriction on the domain of quantification. Maybe when (6) is heard as reporting John's reason for going to the store, we have restricted the domain of our quantifiers so that the only causes we are considering are causes that "correspond" to John's reasons for action. Here's how this might work if RATIONALIZATIONS is true: when we use (6) to report John's reason for acting, the domain is restricted to exclude causes of John's going that do not make the right-hand side of clause (i) of the theory true. (So if the fact that P causes John to go, but that fact does not do this by causing John to believe to P; or this belief does not (possibly with other states) rationalize going; or the fact that John has this rationalization does not also cause him to go; or the rationalization does not cause him go by causing him to manifest a disposition to act on rationalizations; then the fact that P is excluded from the domain of quantification.) The role of (i) in RATIONALIZATIONS, then, is not so much to say anything about the "metaphysics" of reasons for action, as to make explicit the contextual restrictions we put on which reasons are relevant when we talk about reasons for action.

I think that quantifier domain restriction does help make sense of some ways of conveying an agent's reason for acting. I can say "John went to the store because he thought it sold milk" to mean that John's reason for going was that it sold milk. It is plausible that when I say this I have restricted the domain to those causes of John's going that are beliefs of John's, the contents of which give his reasons for going, so that "John's belief that P caused him to go to the store" is true in the context if and only if "One of John's reasons for going was that P" is also true. (I might choose to mention his belief rather than just the belief's content to signal that John was wrong about whether the store sold milk.) But I don't think appealing to quantifier domain restriction works in general to reconcile theories (T0) or (T1f) with the theories of acting for reasons discussed in section 5.2.

I said how the domain restriction strategy is supposed to work for the part of RATIONALIZATIONS that concerns factual reasons for action. How is it supposed to work for the part that concerns non-finite reasons for action? Things here are not as straightforward. If we take TELEOLOGY-1 to be the correct theory of non-finite reasons why (that are not an agent's reason for action), one idea is that in contexts in which we are discussing reasons for action, "the reason why X ϕ-ed was to Z" is true iff the fact that ϕ-ing was a way for X to Z caused X to ϕ, *and some other conditions are met*—those conditions presumably including the requirement that this fact caused X to ϕ by causing him to want to Z. But this will not work, for it can be that the reason why John went to the store was to get some milk, and this is heard as reporting his reason for going to the store, even if the store lacks milk, and so there is no such fact as the fact that going to the store was a way to get milk to cause him to act.[51]

The appeal to quantifier domain restriction does not work even for the part of RATIONALIZATIONS that is about factual reasons. The problem is that "The reason why X ϕ-ed was that P" can be true even when the fact that P does *not* cause X's ϕ-ing. For in cases like this the quantifier domain restriction strategy must say that "The reason why X ϕ-ed was that P" is false—that strategy requires that even in contexts where we are

---

[51] I did suggest that TELEOLOGY-1 needed to be revised in light of the problem of ineffective means. My argument here though is that sometimes people take means that are not just ineffective but (as I said earlier) completely misguided: even under idealized conditions, ϕ-ing is not a way to Z.

talking about reasons for action, reasons are causes (it just denies that, in such contexts, all causes are reasons). Here is a case of the required kind: a mathematics enthusiast goes out celebrating, and when asked why he replies that the reason why he is celebrating is that the 100,000th digit of π is 1. (Maybe he just learned this fact, and thought it was tremendously exciting.) What he says could be true: the sentence

(7)   The main reason why he is celebrating is that the 100,000th digit of π is 1.

has a true reading in the context in which he uttered it. But in no context is "the fact that the 100,000th digit of π is 1 a cause of his celebrating" true.

In this case, the fact that the 100,000th digit of π is prevented from being a cause by the fact that it is wholly about abstract objects. Facts about the future are also unable to be causes, and so are another source of counterexamples. It could be that the reason why John is buying a cake is that there will be a party this weekend. But in no context is "the fact that there will be a party is a cause of John's purchase" true.[52]

It will not do to say that it is false that the reason why the mathematician is celebrating is that the 100,000th digit of π is 1, that the "real" reason why he is celebrating is that *he had just come to believe* that the 100,000th digit of π is 1, at least not if we are using "reason why" so that it reports his reason for celebrating. For if we did, parallel reasoning would require us to say that if Jones, who wants milk, comes to believe that there is milk at the store, and as a result goes to the store intentionally, then Jones's reason for going to the store must be that he came to believe that there is milk at the store. But this needn't be his reason, in fact that would be quite strange. His reason for going could be—and in the normal case is—the content of his belief, not the fact that he came to have a belief with that content.

I have presupposed (T0), that all reasons why something happens are causes of its happening, in presenting this argument. But I think even those who think that there are "non-causal explanations" will agree with my claim that (7) is false when offered as an answer to the question why the mathematician is celebrating, when this question is not heard as a request for his reasons for celebrating.

---

[52] Thanks here to Kieran Setiya.

7

Conclusion

I have argued that a "theory of explanation" should be a theory of answers to why-questions, and that the core of a theory of answers to why-questions should be a theory of reasons why. I have also urged that we should stop arguing about what it takes to be an "explanation," or a "causal explanation." The first term is too ambiguous (between, among other meanings, different kinds of answers), and the second one too vague.

I have offered, and defended at length, a particular theory of reasons why: (T1f), which entails that the only reasons why any event occurs are its causes and its grounds.

My defense of (T1f) leaned heavily on two distinctions.

The first is the distinction between the reasons why E happened, and the reasons why those reasons are reasons. This distinction corresponds to the distinction between the question why E happened, and the question, concerning an answer to this first question, of why that answer is an answer. I claimed that higher-level reasons, reasons why something is a reason why E happened, are not always lower-level reasons, reasons why E happened.

The second distinction is the distinction between an answer to a question, and a good response to that question. Not all good responses to a question contain only an answer to that question; often a good response also contains answers to natural but unasked follow-up questions.

I used these distinctions to argue that many examples people have offered as examples of "non-causal explanations of events," that is, of reasons why some event happened that are not causes (or grounds) of that event, fail to be counterexamples to (T1f). They can look like counterexamples if we take some higher-level reason and assume that it is also a lower-level reason. We can be seduced into doing this if we find the higher-level reason R in a good response to the question why E happened, and assume that anything in a good response is part of the answer, and so that R is a reason why E happened. But we should not assume these things, for they are false. Assuming them can lead us from truths to falsehoods; that is what happens in the examples I discussed; that is why the examples fail to be counterexamples.

There are lots more examples I could have discussed, examples that one or another philosopher has thought were not "causal explanations," including more examples from biology and examples that make use of statistical facts. But my aim was not to show in exhaustive detail how every attempt to refute a theory like (T1f) has failed. I am confident that the distinctions I have used to deal with the examples I have discussed are sufficient to deal with other examples as well.

Despite these successes (I think they are successes), I have not been able to defend in this book a completely general theory of reasons why, an account of what it takes for "that R is a reason why Q" to be true no matter what goes in for "Q," not just when what goes in for "Q" describes the occurrence of an event (though I did offer such a theory, in Appendix B). Nor have I been able to offer a completely unified theory of reasons why events occur. The theory (T1f) is not unified because it is disjunctive. In this book especially, one should ask why it is that all the reasons are either causes or grounds. Does this question have an answer? Or is there nothing to be said about what causation and grounding have in common, in virtue of which they and only they generate reasons why (in the case of events)?

A natural thought, one many others have had, is that there is a connection between dependence, and being a reason why:[1] that R is a reason why Q iff the fact that Q depends on the fact that R. If this is right we could then say that the reason why causes are reasons is that causation is a form of dependence. (This would, then, be a revision of (T1f), which says merely that the main reason why causes are reasons is that they are causes.) We could even say, in general, whether or not the reasons in question are reasons why any event occurred, that the reason why those reasons are reasons why Q is that the fact that Q depends on those reasons.

I don't, however, regard this as much progress, until we have an account of what the relevant notion of dependence is, and a story about why causation and grounding are species of it. Some philosophers have argued

---

[1] Though they articulate it as a connection between dependence and explanation; see for example, Strevens, in *Depth*, section 5.7.

that there are very close analogies between causation and grounding;[2] maybe those analogies are the place to look.

Philosophers of science want to know what it takes to answer a why-question because answering why-questions about the world around us, and about ourselves, is one of the aims of science. Now it is a fact that science does a lot more than just seek out the causes of things. Many scientific research programs—statistical mechanics is an excellent example—aim to discover what grounds the occurrence of certain events, not, or not just, what caused those events.[3] While this is a problem for (T0), it is only to be expected if (T1f) is right. What might be thought a problem for (T1f) is the fact that science also does a lot more than just seek out the causes *and grounds* of events. But I do not see this fact as a problem for (T1f). For not all why-questions ask why some event occurred. And science aims to answer why-questions that are not about events just as much as it aims to answer questions that are. Indeed, as we have seen, pursuing the second aim easily leads to pursuit of the first. Looking for the answer to a why-question that is about an event can easily raise why-questions that aren't. Finding the answer to the question why E happened, figuring out what caused E, raises several related follow-up why-questions: why is that answer the answer? Why did that cause cause its effect?

I do not, however, mean to suggest that the search for second-level reasons is an afterthought, or even that it is a separate project from searching for first-level reasons. The two projects are usually intertwined. In many cases one discovers the first- and second-level reasons simultaneously, and comes to know what the first-level reasons are only by learning why they are reasons. It was obvious that humans in the year 1000 were disposed to have both sons and daughters, but it was far from obvious that this is a cause of humans' now being disposed to have sons and daughters in equal numbers, and far from obvious that it is a reason why humans are so disposed. What it took to know that it is cause, and a reason, was figuring out that the disposition to have sons and daughters in equal numbers is a global equilibrium state. Facts about the structure

---

[2] See, for example, Schaffer, "Grounding in the Image of Causation," and Shaheen, "The Causal Metaphor Account of Metaphysical Explanation." Wilson, in "Metaphysical Causation," argues that grounding is a kind of causation. Bennett, in *Making Things Up,* argues that grounding and causing are both kinds of "building" relations.

[3] Kit Fine wrote that "Ground . . . stands to philosophy as cause stands to science" ("Guide to Ground," p. 40). It's also true that ground stands to philosophy as *ground* stands to science.

of the "space of possible birth-ratio dispositions," including the fact that the equal disposition is an equilibrium, are reasons why the fact about the year 1000 is a cause and a reason. In my view evolutionary theory's great achievement in this case was finding these second-order reasons. A great many scientific achievements consist in finding what I have called higher-level reasons. My achievement, I hope, has been to figure out where those reasons belong.

References

Achinstein, Peter. *The Nature of Explanation*. Oxford University Press, 1983.

Allen, Colin and Mark Bekoff. "Biological Function, Adaptation, and Natural Design." *Philosophy of Science* vol. 62, 1995, 609–22.

Alvarez, M. and Hyman, J. "Agents and their Actions." *Philosophy* vol. 73, 1998, 219–45.

Anscombe, G. E. M. *Intention*. 2nd Edition. Harvard University Press, 1963.

Arpaly, Nomy. *Merit, Meaning, and Human Bondage*. Princeton University Press, 2006.

Arpaly, Nomy and Timothy Schroeder. "A Causal Theory of Acting for Reasons." *American Philosophical Quarterly* vol. 52, 2015, 103–14.

Arpaly, Nomy and Timothy Schroeder. *In Praise of Desire*. Oxford University Press, 2014.

Austin, J. L. *How To Do Things With Words*. 2nd Edition. J. O. Urmson and Marina Sbisà (eds.). Harvard University Press, 1975.

Baker, C. L. *Indirect Questions in English*. PhD Dissertation, University of Illinois at Urbana-Champaign, 1968.

Batterman, Robert. "Multiple Realizability and Universality." *The British Journal for the Philosophy of Science* vol. 51, 2000, 115–45.

Batterman, Robert. "On the Explanatory Role of Mathematics in Empirical Science." *The British Journal for Philosophy of Science* vol. 61, 2010, 1–25.

Bedau, Mark. "Can Biological Teleology Be Naturalized?" *The Journal of Philosophy* vol. 88, 1991, 647–55.

Bedau, Mark. "Where's the Good in Teleology?" *Philosophy and Phenomenological Research* vol. 52, 1992, 781–806.

Beebee, Helen. "Causing and Nothingness." In John Collins, Ned Hall, and L. A. Paul (eds.), *Causation and Counterfactuals*. Oxford University Press, 2004, 291–308.

Bennett, Jonathan. *Events and Their Names*. Hackett, 1988.

Bennett, Karen. *Making Things Up*. Oxford University Press, Forthcoming.

Boorse, Christopher. "Wright on Functions." *The Philosophical Review* vol. 85, 1976, 70–86.

Braine, David. "Varieties of Necessity." *Proceedings of the Aristotelian Society* vol. 46, 1972, 139–87.

Brandon, Robert N. "The Principle of Drift: Biology's First Law." *The Journal of Philosophy* vol. 103, 2006, 319–35.

Bratman, Michael. "Davidson's Theory of Intention." In B. Vermazen and M. Hintikka (eds.), *Essays on Davidson: Actions and Events*. Oxford University Press, 1985, 13–26.

Bromberger, Sylvain. "An Approach to Explanation." In *On What We Know We Don't Know*, 18–51.

Bromberger, Sylvain. *On What We Know We Don't Know*. The University of Chicago Press and CSLI, 1992.

Bromberger, Sylvain. "Why-Questions." In *On What We Know We Don't Know*, 75–100.

Broome, John. *Rationality Through Reasoning*. Wiley-Blackwell, 2013.

Cohen, G. A. *Karl Marx's Theory of History: A Defence*. Princeton University Press, 1978.

Colyvan, Mark. *The Indispensability of Mathematics*. Oxford University Press, 2001.

Coope, Ursula. "Aristotle on Action." *Proceedings of the Aristotelian Society* vol. 81, 2007, 109–38.

Cooper, John. "Aristotle on Natural Teleology." In *Knowledge, Nature, and the Good: Essays on Ancient Philosophy*. Princeton University Press, 2004.

Cross, Charles. "Explanation and the Theory of Questions." *Erkenntnis* vol. 34, 1991, 237–60.

Dancy, Jonathan. *Ethics Without Principles*. Oxford University Press, 2004.

Davidson, Donald. "Actions, Reasons, and Causes." In *Essays on Actions and Events*, 3–20.

Davidson, Donald. "Causal Relations." *The Journal of Philosophy* vol. 64, 1967, 691–703.

Davidson, Donald. *Essays on Actions and Events*. 2nd Edition. Oxford University Press, 2001.

Davidson, Donald. "Freedom to Act." In *Essays on Actions and Events*, 63–82.

Davidson, Donald. "On the Very Idea of a Conceptual Scheme." In *Inquiries into Truth and Interpretation*. 2nd Edition. Oxford University Press, 2001, 183–98.

Des Chene, Dennis. *Physiologia: Natural Philosophy in Late Aristotelian and Cartesian Thought*. Cornell University Press, 1996.

Dretske, Fred. *Explaining Behavior: Reasons in a World of Causes*. MIT Press, 1988.

Dretske, Fred. "Triggering and Structuring Causes." In Timothy O'Connor and Constantine Sandis (eds.), *A Companion to the Philosophy of Action*. Blackwell, 2010.

Elga, Adam. "Isolation and Folk Physics." In Huw Price and Richard Corry (eds.), *Causation, Physics, and the Constitution of Reality*, Oxford University Press, 2007, 106–19.

Fara, Michael and Sungho Choi. "Dispositions." In Edward N. Zalta (ed.), *The Stanford Encyclopedia of Philosophy* (Spring 2014 Edition), URL = <http://plato. stanford.edu/archives/spr2014/entries/dispositions/>.

Field, Hartry. "Causation in a Physical World." In Michael J. Loux and Dean W. Zimmerman (eds.), *The Oxford Handbook of Metaphysics*. Oxford University Press, 2003, 435–60.

Fine, Kit. "Guide to Ground." In Fabrice Correia and Benjamin Schnieder (eds.), *Metaphysical Grounding*. Cambridge University Press, 2012.

Fisher, R. A. *The Genetical Theory of Natural Selection*. Dover Publications, 1958.

Frisch, Mathias. *Causal Reasoning in Physics*. Cambridge University Press, 2014.

Garfinkel, Alan. *Forms of Explanation*. Yale University Press, 1981.

Grice, Paul. "Logic and Conversation." In *Studies in the Ways of Words*. Harvard University Press, 1991, 22–40.

Groenendijk, J. and M. Stokhof. "Semantic Analysis of Wh-Complements." *Linguistics and Philosophy* vol. 5, 1982, 175–233.

Gumbel, Andrew. "Oklahoma City Bombing: 20 years later, key questions remain unanswered." *theguardian.com*, April 13, 2015. URL = <http://www.theguardian.com/us-news/2015/apr/13/oklahoma-city-bombing-20-years-later-key-questions-remain-unanswered>. Accessed April 14, 2015.

Halpern, J. and J. Pearl. *Causes and Explanations: A Structural Model Approach*. Technical report R-266, Cognitive Systems Laboratory. University of California (Los Angeles), 2000.

Hamblin, C. L. "Questions." *Australasian Journal of Philosophy* vol. 36, 1958, 159–68.

Hamblin, C. L. "Questions in Montague English." *Foundations of Language* vol. 10, 1973, 41–53.

Hawthorne, John. "Before-Effect and Zeno Causality." *Nous* vol. 34, 2000, 622–33.

Hawthorne, John and David Manley. Review of Mumford's *Dispositions*. *Nous* vol. 39, 2005, 179–95.

Hawthorne, John and Daniel Nolan. "What Would Teleological Causation Be?" In John Hawthorne, *Metaphysical Essays*. Oxford University Press, 2006, 264–84.

Heim, Irene. "Concealed Questions." In R. Baüerle, U. Egli, and A. von Stechow (eds.), *Semantics from Different Points of View*. Springer-Verlag, 1979, 51–60.

Hempel, Carl. "Aspects of Scientific Explanation." In *Aspects of Scientific Explanation and Other Essays in the Philosophy of Science*. The Free Press, 1965, 331–496.

Hempel, Carl. "Deductive-Nomological vs. Statistical Explanation." In Herbert Feigl and Grover Maxwell (eds.), *Minnesota Studies in the Philosophy of Science III*. University of Minnesota Press, 1962.

Hempel, Carl and Paul Oppenheim. "Studies in the Logic of Explanation." *Philosophy of Science* vol. 15, 1948, 135–75.

Hitchcock, Christopher. "Farewell to Binary Causation." *Canadian Journal of Philosophy* vol. 26, 1996, 267–82.

Hitchcock, Christopher. "Probabilistic Causation." In Edward N. Zalta (ed.), *The Stanford Encyclopedia of Philosophy* (Winter 2012 Edition), URL = <http://plato.stanford.edu/archives/win2012/entries/causation-probabilistic/>.

Hitchcock, Christopher. "The Intransitivity of Causation Revealed in Equations and Graphs." *The Journal of Philosophy* vol. 98, 2001, 273–99.

Hitchcock, Christopher. "The Role of Contrast in Causal and Explanatory Claims." *Synthese* vol. 103, 1996, 395–419.

Huddleston, Rodney and Geoffrey K. Pullum. *A Student's Introduction to English Grammar*. Cambridge University Press, 2005.

Huddleston, Rodney and Geoffrey K. Pullum. *The Cambridge Grammar of the English Language*. Cambridge University Press, 2002.

Jackson, Frank and Philip Pettit. "In Defense of Explanatory Ecumenism." *Economics and Philosophy* vol. 8, 1992, 1–21.

Karttunen, Lauri. "Syntax and Semantics of Questions." *Linguistics and Philosophy* vol. 1, 1977, 3–44.

Kim, Jaegwon. "Hempel, Explanation, Metaphysics." *Philosophical Studies* vol. 94, 1999, 1–20.

Kitcher, Philip. "Explanatory Unification." *Philosophy of Science* vol. 48, 1981, 507–31.

Kment, Boris. "Counterfactuals and Explanation." *Mind* vol. 115, 2006, 261–309.

Lahiri, Utpal. *Questions and Answers in Embedded Contexts*. Oxford University Press, 2002.

Lange, Marc. "Laws and Meta-laws of Nature: Conservation Laws and Symmetries." *Studies in History and Philosophy of Modern Physics* vol. 38, 2007, 457–81.

Lange, Marc. "What Makes a Scientific Explanation Distinctively Mathematical?" *The British Journal for the Philosophy of Science* vol. 64, 2013, 485–511.

Lenman, James. "Reasons for Action: Justification vs. Explanation." In Edward N. Zalta (ed.), *The Stanford Encyclopedia of Philosophy* (Winter 2011 Edition), URL = <http://plato.stanford.edu/archives/win2011/entries/reasons-just-vs-expl/>.

Leuenberger, Stephan. "Grounding and Necessity." *Inquiry* vol. 57, 2014, 151–74.

Levine, Joseph. "Materialism and Qualia: The Explanatory Gap." *Pacific Philosophical Quarterly* vol. 64, 1983, 354–61.

Lewis, David. "Causal Explanation." In *Philosophical Papers Volume II*. Oxford University Press, 1986, 214–40.

Lewis, David. "Events." In *Philosophical Papers Volume II*. Oxford University Press, 1986, 241–69.

Lipton, Peter. "Understanding without Explanation." In Hank W. de Regt, Sabina Leonelli, and Kai Eigner (eds.), *Scientific Understanding: Philosophical Perspectives*. University of Pittsburgh Press, 2009.

Manning, Richard N. "Biological Function, Selection, and Reduction." *The British Journal for the Philosophy of Science* vol. 48, 1997, 69–82.

Maudlin, Tim. "Causes, Counterfactuals, and the Third Factor." In John Collins, Ned Hall, and L. A. Paul (eds.), *Causation and Counterfactuals*. Oxford University Press, 2004, 419–44.

Mellor, D. H. "For Facts as Causes and Effects." In John Collins, Ned Hall, and L. A. Paul (eds.), *Causation and Counterfactuals*. Oxford University Press, 2004, 309–23.

Nagel, Thomas. *The Possibility of Altruism*. Oxford University Press, 1970.

Neander, Karen. "The Teleological Notion of 'Function.'" *Australasian Journal of Philosophy* vol. 69, 1991, 454–68.

Nerlich, Graham. "What Can Geometry Explain?" *The British Journal for the Philosophy of Science* vol. 30, 1979, 69–83.

Paul, L. A. and Ned Hall. *Causation: A User's Guide*. Oxford University Press, 2013.

Peacocke, Christopher. *Holistic Explanation*. Oxford University Press, 1979.

Pearl, Judea. *Causality: Models, Reasoning and Inference*. Cambridge University Press, 2000.

Pincock, Christopher. "A Role for Mathematics in the Physical Sciences." *Nous* vol. 42, 2007, 253–75.

Price, Huw and Richard Corry (eds.). *Causation, Physics, and the Constitution of Reality: Russell's Republic Revisited*. Oxford University Press, 2007.

Railton, Peter. "A Deductive-Nomological Model of Probabilistic Explanation." *Philosophy of Science* vol. 45, 1978, 206–26.

Railton, Peter. "Probability, Explanation, and Information." *Synthese* vol. 48, 1981, 233–56.

Rosen, Gideon. "Metaphysical Dependence: Grounding and Reduction." In Bob Hale and Aviv Hoffmann (eds.), *Modality: Metaphysics, Logic, and Epistemology*. Oxford University Press, 2010, 109–36.

Russell, Bertrand. "On the Notion of Cause." *Proceedings of the Aristotelian Society* (New Series) vol. 13, 1912/13, 1–26.

Russell, E. S. *The Directiveness of Organic Activities*. Cambridge University Press, 1945.

Salmon, Wesley. "A Third Dogma of Empiricism." Reprinted in *Causality and Explanation*. Oxford University Press, 1998, 95–107.

Salmon, Wesley. *Four Decades of Scientific Explanation*. University of Pittsburgh Press, 1989.

Salmon, Wesley. *Scientific Explanation and the Causal Structure of the World*. Princeton University Press, 1984.

Salmon, Wesley. *Statistical Explanation and Statistical Relevance*. With contributions by J. G. Greeno and R. C. Jeffrey. University of Pittsburgh Press, 1971.

Schaffer, Jonathan. "Contrastive Causation." *The Philosophical Review* vol. 114, 2005, 297–328.

Schaffer, Jonathan. "Grounding in the Image of Causation." *Philosophical Studies*, forthcoming.

Scheffler, Israel. "Thoughts on Teleology." *The British Journal for the Philosophy of Science* vol. 9, 1959, 265–84.

Schlosser, Markus E. "Basic Deviance Reconsidered." *Analysis* vol. 67, 2007, 186–94.

Schnieder, Benjamin. "A Logic for 'Because'." *The Review of Symbolic Logic* vol. 4, 2011, 445–65.

Schroeder, Mark. *Slaves of the Passions*. Oxford University Press, 2007.

Scriven, Michael. "Explanations, Predictions, and Laws." In Herbert Feigl and Grover Maxwell (eds.), *Minnesota Studies in the Philosophy of Science III*. University of Minnesota Press, 1962, 170–230.

Scriven, Michael. "Truisms as the Grounds for Historical Explanation." In P. Gardiner (ed.), *Theories of History*. The Free Press, 1959, 443–75.

Setiya, Kieran. *Knowing Right from Wrong*. Oxford University Press, 2012.

Setiya, Kieran. "Practical Knowlege." *Ethics* vol. 118, 2008, 388–409.

Setiya, Kieran. "Reasons and Causes." *European Journal of Philosophy* vol. 19, 2009, 129–57.

Setiya, Kieran. *Reasons Without Rationalism*. Princeton University Press, 2007.

Setiya, Kieran. "Sympathy for the Devil." In Sergio Tenenbaum (ed.), *Desire, Practical Reason, and the Good*. Oxford University Press, 2010, 82–110.

Setiya, Kieran. "What is a Reason to Act?" *Philosophical Studies* vol. 167, 2014, 221–35.

Shaheen, Jonathan. *Meaning and Explanation*. PhD Dissertation, University of Michigan, 2015.

Shaheen, Jonathan. "The Causal Metaphor Account of Metaphysical Explanation." Unpublished.

Skiles, Alexander. "Against Grounding Necessitarianism." *Erkenntnis* vol. 80, 2015, 717–51.

Skow, Bradford. "Are Shapes Intrinsic?" *Philosophical Studies* vol. 133, 2007, 111–30.

Skow, Bradford. "Are There Genuine Physical Explanations of Mathematical Phenomena?" *The British Journal for the Philosophy of Science* vol. 66, 2015, 69–93.

Skow, Bradford. "Are There Non-Causal Explanations (of Particular Events)?" *The British Journal for the Philosophy of Science* vol. 65, 2014, 445–67.

Sober, Elliott. *Did Darwin Write the Origin Backwards?* Prometheus Books, 2011.

Sober, Elliott. "Equilibrium Explanation." *Philosophical Studies* vol. 43, 1983, 201–10.

Sober, Elliott. *The Nature of Selection*. University of Chicago Press, 1984.

Spirtes, Peter, Clark Glymour, and Richard Scheines. *Causation, Prediction, and Search*. The MIT Press, 2000.

Stalnaker, Robert. "Pragmatic Presuppositions." In *Context and Content*. Oxford University Press, 1999, 47–62.

Stanley, Jason. *Know How*. Oxford University Press, 2011.

Stanley, Jason, and Zoltán Gendler Szabó. "On Quantifier Domain Restriction." *Mind & Language* vol. 15, 2000, 219–61.

Strevens, Michael. *Depth*. Harvard University Press, 2008.

Taylor, Charles. *The Explanation of Behavior*. Prometheus Books, 1964.

Taylor, Richard. "Purposeful and Non-Purposeful Behavior: A Rejoinder." *Philosophy of Science* vol. 17, 1950, 327–32.

Thompson, Michael. *Life and Action*. Harvard University Press, 2008.

Toulmin, Stephen. *An Examination of the Place of Reason in Ethics*. Cambridge University Press, 1950.

Van Fraassen, Bas C. *The Scientific Image*. Oxford University Press, 1980.

Van Inwagen, Peter. "Causation and the Mental." In *Existence: Essays in Ontology*. Cambridge University Press, 2014.

Wedgwood, Ralph. "The Normative Force of Reasoning." *Nous* vol. 40, 2006, 660–86.

Weslake, Brad. "Proportionality, Contrast and Explanation." *Australasian Journal of Philosophy* vol. 91, 2013, 785–97.

Wilson, Alastair. "Metaphysical Causation." Unpublished.

Woodfield, Andrew. *Teleology*. Cambridge University Press, 1976.

Woodward, James. *Making Things Happen*. Oxford University Press, 2003.

Woodward, James and Christopher Hitchcock. "Explanatory Generalizations, Part I: A Counterfactual Account." *Nous* vol. 37, 2003, 1–24.

Wright, Larry. "The Case against Teleological Reductionism." *The British Journal for the Philosophy of Science* vol. 19, 1968, 211–23.

Wright, Larry. *Teleological Explanations*. University of California Press, 1976.

Yablo, Stephen. "Advertisement for a Sketch of an Outline of a Proto-theory of Causation." In *Things*. Oxford University Press, 2010, 98–116.

Yablo, Stephen. "Mental Causation." In *Thoughts*. Oxford University Press, 2008, 222–48.

Yablo, Stephen. "Wide Causation." In *Thoughts*. Oxford University Press, 2008, 275–306.

Index